AF744048

Urban Health and Wellbeing

Systems Approaches

Series Editor

Yongguan Zhu, Chinese Academy of Sciences, Institute of Urban Environment, Xiamen, Fujian, China

The Urban Health and Wellbeing: Systems Approaches series is based on a 10-year global interdisciplinary research program developed by International Council for Science (ICSU), and sponsored by the InterAcademy Partnership (IAP) and the United Nations University (UNU). It addresses up-to-date urban health issues from around the world and provides an appealing integrated urban development approach from a systems perspective. This series aims to propose a new conceptual framework for considering the multi-factorial and cross sectorial nature of both determinants and drivers of health and wellbeing in urban populations and takes a systems approach for improving the understanding of the interconnected nature of health in cities. The systems approach includes an engagement with urban communities in the process of creating and transferring knowledge. Further, it aims at generating knowledge and providing the evidence that is relevant to people and policy-makers for improving integrated decision making and governance for the health and wellbeing of urban dwellers. The methods applied, come from various epistemological domains in order to improve understanding of how the composition and functioning of urban environments impacts physical, mental and social heath and how inequalities can be reduced to improve the overall quality of urban life.

The systems approach is applied to science and society and defined by a deep investigation into disciplinary knowledge domains relevant for urban health and well-being, as well as an inter- and transdisciplinary dialogue and shared understanding of the issues between scientific communities, policy makers and societal stakeholders more broadly. It involves one or more of the following elements: 1) the development of new conceptual models that incorporate dynamic relations among variables which define urban health and wellbeing; 2) the use of systems tools, stimulation models and collaborative modelling methods; 3) the integration of various sources and types of data including spatial, visual, quantitative and qualitative data.

Like the first book, the coming books will all address the topic of urban health and wellbeing, specifically by taking a systems approach. The topics range across all urban sectors and can, for example, cover the following areas:

(1) transportation, urban planning and housing, urban water, energy and food, communication, resources and energy, urban food systems, public service provision, etc.
(2) the related health disorders in physical, social and mental health
(3) the methods and models used and the type of science applied to understand the complexity of urban health and wellbeing.

Jieling Liu

Transforming Urban Green Space Governance in China Under Ecological Civilization: An Institutional Analysis

ZHEJIANG UNIVERSITY PRESS
浙江大学出版社

Springer

Jieling Liu
Institute of Social Sciences
University of Lisbon
Lisbon, Portugal

ISSN 2510-3490 ISSN 2510-3504 (electronic)
Urban Health and Wellbeing
ISBN 978-981-99-6693-6 ISBN 978-981-99-6694-3 (eBook)
https://doi.org/10.1007/978-981-99-6694-3

Jointly published with Zhejiang University Press
The print edition is not for sale in China (Mainland). Customers from China (Mainland) please order the print book from: Zhejiang University Press.

This research is funded by the Portuguese National Foundation for Science and Technology (FCT).

© Zhejiang University Press 2023, corrected publication 2025

This work is subject to copyright. All rights are solely and exclusively licensed by the Publisher, whether the whole or part of the material is concerned, specifically the rights of reprinting, reuse of illustrations, recitation, broadcasting, reproduction on microfilms or in any other physical way, and transmission or information storage and retrieval, electronic adaptation, computer software, or by similar or dissimilar methodology now known or hereafter developed.
The use of general descriptive names, registered names, trademarks, service marks, etc. in this publication does not imply, even in the absence of a specific statement, that such names are exempt from the relevant protective laws and regulations and therefore free for general use.
The publishers, the authors, and the editors are safe to assume that the advice and information in this book are believed to be true and accurate at the date of publication. Neither the publishers nor the authors or the editors give a warranty, expressed or implied, with respect to the material contained herein or for any errors or omissions that may have been made. The publishers remain neutral with regard to jurisdictional claims in published maps and institutional affiliations.

This Springer imprint is published by the registered company Springer Nature Singapore Pte Ltd.
The registered company address is: 152 Beach Road, #21-01/04 Gateway East, Singapore 189721, Singapore

Paper in this product is recyclable.

This book is edited from the PhD Thesis which the main author, Jieling Liu, had developed for obtaining the doctoral degree in *Climate Change and Sustainable Development Policies*, specialty in *Geography and Planning*.

勿忘初心，为社会和环境作贡献。

The original version of the book has been revised: Author name Franz Gatzweiler has been removed from this title. A correction to this book can be found at https://doi.org/10.1007/978-981-99-6694-3_9

Acknowledgements

Undertaking this Ph.D. has been a life-transforming experience for me, beyond doubt, and it would not have been possible to achieve without the generous support of many professors, friends, family members, and organisations. I would like to acknowledge a few.

First and foremost, I am profoundly grateful to my two supervisors for their tireless efforts to cultivate my abilities towards a Ph.D. My principal supervisor, Prof. Franz Gatzweiler, has supported me in every way he could and tirelessly so on research advice, opportunities, research resources, and inspired my thinking on every important subject matter related to this book. He has frequently guided and advised me on the research subject and encouraged me to present papers at conferences and to integrate into the professional networks in my research area. His insightful feedback during our discussions pushed me to sharpen my thinking and brought my work to a higher level. Besides valuable input, he also set a role model for outstanding research, scholarship, and mentorship. My co-supervisor, Prof. Olivia Bina, has contributed much valuable advice and feedback on the research topic, direction, and the analytical aspects. She has constantly supported me by monitoring the progress and making sure that I am aware of the evaluation requirements and their periods and deadlines to progress in the Ph.D. She has also been very patient and encouraged me to focus on completing this book and with good quality.

I would also like to acknowledge the many admirable professors who diligently equipped me with the essential knowledge foundation and research skills during the first year of the Ph.D., including Prof. Luísa Schmidt, Prof. Filipe Duarte Santos, Prof. Viriato Soromenho-Marques, Prof. Júlia Seixas, Prof. Rui Ferreira Santos, Prof. Ana Delicado, Prof. Mónica Trunninger, Prof. Pedro Prista Monteiro, Prof. Gil Penha-Lopes, Prof. José Saldanha Matos, Prof. João Ferrão, Prof. João Mourato, Prof. José Lima Santos, Prof. Rodrigo Proença de Oliveira, and Prof. Tim O'Riordan. It was a very inspiring, mind-opening, and intellectually pleasing year for me. Besides, I would like to thank Prof. Mário Baptista Coelho and Prof. Miguel Brito at the Faculty of Sciences—University of Lisbon as well for preparing me to enter the Ph.D. programme with critical insights on renewable energies.

Some special words of gratitude go to my host and colleagues at the Ostrom Workshop, where I gained substantial theoretical knowledge about institutions and governance as a visiting fellow. They are Prof. Burnell Fischer (my host), Prof. William Blomquist, Prof. Scott Shackelford, Prof. Erik Nordman, Dr. Ruta Śpiewak, Prof. Eduardo Brondizio, Dr. Hita Unnikrishnan, Prof. Stefan Kolev, Dr. Joe Lambke, Gayle Higgins, Robin Humphrey, David Price, Emily Castle, and Allison Sturgeon. My gratitude naturally includes the legendary Prof. Elinor Ostrom, who I will never have the opportunity to meet in person, but in many ways, the interactions I had with her through reading, digesting, and Ecosystem-based Adaptation her work with colleagues have felt very personal and inspiring.

I am also grateful for Prof. Martin de Jong, Prof. Kongjian Yu, Prof. Peter Ho, and Prof. Rong Tan for providing me valuable research advice throughout the Ph.D. I would also like to announce my appreciation towards all those interviewees who have accepted to work with me on this research or facilitated my access into critical policy archives, without getting nor even expecting rewards. Unfortunately, I cannot name them here individually due to privacy reasons. Their support has been truly essential for the success of this research. I want to thank Dr. Marcus Grant for giving me the experience of being a Lead Guest Editor for a special issue at *Cities and Health*. I am grateful to Dr. Nicholas You, a veteran urban specialist, and the Executive Director of the Guangzhou Institute for Urban Innovation, who inspired me a great deal. I also appreciate Dr. Peter Head, Dr. Stephen Passmore, Dr. José Siri, Dr. Charles Ebikeme, Dr. Manasi Kumar, Prof. Cunrui Huang, Dr. Shaily Gandhi, and Dr. Yash Shukla for the inspiring conversations and collaborations we shared.

My gratitude also goes to the Portuguese Foundation for Science and Technology (Fundação para a Ciência e Tecnologia, FCT), Institute of Urban Environment—Chinese Academy of Sciences (IUE-CAS), and Ostrom Workshop—Indiana University Bloomington. These organisations have provided me with generous scholarships and stipends, which were essential for me to manage basic needs and focus on intellectual inquiry. I could not thank enough my home institute—the Institute of Social Sciences—University of Lisbon (ICS-ULisboa), which provided me with an amazing, intellectually highly enabling academic environment with its regular high-level international seminars and workshops, well-developed library, cozy café, spacious computer rooms, quiet study rooms and free printing and fast internet services. My appreciation also goes to all the hardworking staffs at ICS, particularly Dr. Maria Goretti Matias and Mrs. Raquel Brito, for their patient and warm support for me to navigate the administrative process of the Ph.D. smoothly.

I would also like to thank my close friends from whom I have been so lucky to meet with and learn, some of whom are also colleagues of the Ph.D. programme: Leon White, Jorge Conesa, Filipe Lisboa, Fronika de Wit, Jiesper Tristan Petersen, Pedro Nunes, Sofia Castelo, Tainan Messina, Patricia Silvério, Roy Bishuajit, Tomás Calheiros, Pedro Macedo, Ping Shou, Vickey Shan Zhang, Apple Ying Tan, Rongyuan Chen, Yongxian Gan, Jianqi Luo, Ruby Tiantian Gui, Mengmeng Cui, Ishtiak Zaman, Tingting Wu, Adnan Arshad, Lurdes Ferreira, Rui Mira, Ali Hamidian, Geoffery Larson, Dapith Iglesias Sanders, Dr. Daniel Auer, Drew Du Bois, José Pedro Rodrigues, and my exceptional sweet four-leg friend, Goji. I am grateful for the constant warm emotional support they have given me. After the book submission leading to the dissertation, Ms. Nazanin Bidabadi and Prof. Rui Barros from the University of Porto accompanied and encouraged me to a great extent—their gentle support and care moved me very much.

Lastly, my sincere acknowledgements go to my family, my mother Chaoqiong Long, my brother Junyu Liu, my father Qiongwen Liu, my cousin Shimin Long, and most importantly, my former partner Nir Aviram and his family, for all their unconditional love, tireless encouragement, and constant presence in my life.

Contents

Acronyms and Abbreviations

CCICED	China Council for International Cooperation on Environment and Development
COP	Conference of the parties
CPR	Common-pool resources
DACCS	Direct air carbon capture and storage
EA	Ecological asset
EBA	Ecosystem-based Adaptation
ECV	Ecological culture services value
EGCA	European Green Capital Award
EPV	Ecosystem provision value
ERV	Ecological regulation services value
FYP	Five-Year Plan (China's national social and economic development plan)
GD-HK-MO GBA	Guangdong-Hong Kong-Macau Greater Bay Area
GDP	Gross domestic product
GEP	Gross Ecosystem Product
GHG	Greenhouse gases
GNH	Gross National Happiness
GPCDCP	Guangdong Provincial Centre for Disease Control and Prevention
HDI	Human Development Index
HiAP	Health in All Policies
HWP	Haizhu Wetland Park
IAD	Institutional Analysis and Development
IASS	Potsdam Institute of Advanced Sustainability Studies
IPBES	Intergovernmental Science-Policy Platform on Biodiversity and Ecosystem Services
IUCN	The International Union for Conservation of Nature
LTPR	Land Tenure and Property Rights
MCA	Multi-criteria assessment
MCE	Multi-criteria evaluation

MEA	Millennium Ecosystem Assessment
MEP	Ministry of Environmental Protection
MOHURD	Ministry of Housing and Urban-Rural Development
MSL	Mean sea level
NBS	Nature-based Solutions
NDCs	Nationally Determined Contributions
NDRC	National Development and Reform Commission
NIE	New Institutional Economics
NUA	New Urban Agenda
PEC	Panyu Ecological Corridor
PRD	Pearl River Delta
QDA	Qualitative Data Analysis
SDG 3	Health and Well-being
SDG 11	Sustainable Cities and Communities
SDG 13	Climate Action
SDGs	Sustainable Development Goals
SEA	Strategic Environmental Assessment
SEEA	UN System of Environmental-Economic Accounting
SEPA	State Environmental Protection Agency
TEEB	Economics of Ecosystems and Biodiversity initiative
TSCD	Tianhe Sponge City demonstration site
UCLG	United Cities and Local Governments
UGS	Urban Green Spaces
ULT	Urban Land Teleconnections
UNFCCC	United Nations Framework Convention on Climate Change
WAVES	World Bank Wealth Accounting and the Valuation of Ecosystem Services
WHO	World Health Organization
WWF	World Wildlife Fund

List of Figures

List of Tables

Chapter 1
Introduction

1.1 Urbanisation Trend, Environmental Degradation, and Climate Change Impacts in China

The global urban population has grown rapidly since 1950, having raised from 30 to 55% in 2018 (UN DESA 2018). In China, the urbanisation rate was even more extraordinary. It grew rapidly from 17.9% in 1978 to 53.7% in 2014; during this period, 560 million population have moved into cities and 465 new cities have been built (State Council of China 2014). The country is projected to become 70% urbanised, with over one billion people living in cities by 2030 (Pan 2016).

The rapid urbanisation in China is considered as one of the biggest human settlement challenges in history, accompanied by profound social, economic, and environmental transformations (Bai et al. 2014; Bina 2008). Urban land-use and landscape patterns have changed as a result of rapid urban development and total population increase (Dorning et al. 2015; Wu 2010). In many emerging urban areas in China, land was used for urban development much faster than population growth. Researchers (Bai et al. 2014) found the area of developed land in China grew by 78.5% between 2003 and 2013, while urban population grew only 46% during this period. Another research reveals approximately 27% of total agricultural land, forests, and grasslands in China being converted to urban areas between 1979 and 2010 (Ren et al. 2013). The remarkable transformation of green spaces to built-up areas has even gained a reference as "landscape urbanisation," in contrast to "demographic urbanisation" (Bai et al. 2012a, b). Future land-use forecasts suggest even more dramatic increases in natural-to-urban land conversion, in other words, land in cities will be increasingly scarce, and more nature areas will be acquired by cities as they continue to grow. Researchers have forecasted a 110–185% of increase in global urban land cover by 2030, with Asia accounting for nearly half of this increase (Angel et al. 2011; Seto et al. 2012).

Unplanned rapid urbanisation, demographic change, and drastic climatic changes are the primary drivers of disaster risks (UNDRR 2015). In China, rapid increases in

© Zhejiang University Press 2023

J. Liu, *Transforming Urban Green Space Governance in China Under Ecological Civilization: An Institutional Analysis*, Urban Health and Wellbeing,
https://doi.org/10.1007/978-981-99-6694-3_1

total population and urbanisation in recent decades have induced significant changes in land-use and landscape patterns, as well as a series of ecological and environmental problems (Gong et al. 2018; Li et al. 2013). Research shows that nearly half of the total urban environment in all city regions in China today, including surface water, air, and soil quality, has been degraded (He et al. 2017).

In addition, changes in climate in recent decades have caused negative impacts on natural and human systems globally (IPCC 2014). These impacts include higher average surface temperatures, higher possibility of extreme events or hazards, more frequent and intense tropical cyclones, inland and coastal flooding, landslides, drought, increased aridity, water scarcity, air pollution, the continuously rising sea levels, storm surges, and fatal heatwaves (Elsner et al. 2008; IPCC 2013; 2014; Walsh et al. 2014).

Many of these climate impacts are concentrated in urban areas, jeopardising human health and well-being, livelihoods, and infrastructure (IPCC 2014). Heatwaves, for example, impose both physical and mental danger, such as heatstroke, depression, and anxiety (Berry et al. 2010; Haines and Patz 2004), particularly on vulnerable communities. In China, heatwaves have significantly increased mortality risk across regions, particularly on densely populated urban communities (Ma et al. 2015). Research shows an evident increase in heatwaves and high-temperature days in Guangdong province in the last five decades, in which Guangzhou City is the capital (Du 2011).

Another negative climate impact on health is air pollution, primarily induced by transport, energy, and industrial production systems that rely on fossil fuels, and occasionally aggravated by wildfires nearby urban settlements that release particulate matter and other toxic substances (Finlay et al. 2012), e.g., the air quality in much part of California in summer 2020. In 2017, the *Lancet Commission on Pollution and Health* report declared air pollution as "the largest environmental cause of disease and death in the world today." According to this report, air pollution was responsible for an estimated 9 million premature deaths globally, that is, 16% of all deaths, in 2015 alone (Landrigan et al. 2018). Many Chinese provinces have an increasing mortality attributed to air pollution in the last three decades. Despite ambitious policy effort in recent years dedicated to reduce industrial air pollution, $PM_{2.5}$ concentrations across all regions still exceed the WHO Air Quality Guideline[1] for the entire population (Yin et al. 2020).

In Guangzhou, a fast-paced Southeast megacity with 16 million urban population,[2] ambient O_3 and NO_2 are the main drivers for increased mortality rate (Wu et al. 2019). A study in 30 Chinese provinces shows that air pollution in 2007 affected 72 million labour forces and caused 346 billion Yuan losses, which equals approximately 1.1% of the national GDP (Xia et al. 2016). Similarly in Europe, pollution from heavy

[1] WHO guidelines for indoor air quality: selected pollutants. Available at: https://www.euro.who.int/en/health-topics/environment-and-health/air-quality/publications/2010/who-guidelines-for-indoor-air-quality-selected-pollutants.

[2] Bureau of Statistics of Guangzhou: Population size and distribution of Guangzhou in 2019 (in Chinese: 2019年广州市人口规模及分布情况). Available at: http://tjj.gz.gov.cn/gkmlpt/content/5/5727/post_5727607.html#226.

goods vehicles alone resulted in 43–46 billion euros of health costs annually (EEA Release 2013). In the future, premature mortality by climate-attributed air pollution is expected to increase greatly, particularly in India and East Asia (Silva et al. 2017). The high physical and financial costs of air pollution make it a pressing issue for many urban governments.

Furthermore, rising sea levels and more frequent and intense tropical cyclones as a result of climate change negatively impact urban population and infrastructure (Elsner et al. 2008; Walsh et al. 2014). Floods and windstorms can adversely affect health, e.g., in the forms of drowning, injuries, hypothermia, and infectious diseases, and especially dengue fever (Schnitzler et al., 2007; Jakubicka et al. 2010). For example, Guangzhou, the megacity located on the low-laying estuary of the Pearl River Delta in Southeast China, is ranked as the fourth city most at risk in terms of population exposed to flooding and storm surge (Hallegatte et al., 2013a; Hanson et al. 2011; IPCC 2014).

In addition to physical health, the increased intensities of the strongest tropical cyclones and the rising sea levels globally continue to impact mental health, leading to fear and stress over exposure to high disaster risks and post-disaster traumas, or the exact opposite—denial, paralysis, and apathy (Berry et al. 2010; Fritze et al. 2008; Swim et al. 2017).

The risks also extend to the urban ecosystems. The observed urbanisation process accompanied by massive landscape and socio-economic changes under climate change is also changing surface temperature and precipitation patterns, soil moisture levels, vegetation growth rates, water and aquifer levels, and air quality, thereby worsening the ecological conditions of cities (Kuttler 2008; Rees and Wackernagel 2008; Revi et al. 2014).

With this ongoing dynamics of change, coastal cities and financially mal-prepared city regions are especially vulnerable (Malik et al. 2012), and most of these city regions are from Asia. A global ranking of 10 megacities[3] with regard to property and infrastructure assets exposed to sea-level rise and coastal flood risks showed eight megacities from Asia: Guangzhou, Kolkata, Shanghai, Mumbai, Tianjin, Tokyo, Hong Kong, and Bangkok; notably, Guangzhou was the most vulnerable (Hallegatte et al. 2013a; Nicholls et al. 2007).

Cities broadly suffer from disaster-related life and economic loss due to climate change impacts; however, cities in developing countries often are particularly at risk due to their substantially higher population density and much more dynamic socio-economic development profiles (Kraas 2008). On this note, experts in climate change and urban health have urged for urgent, cross-sectoral, multi-level collaborative policies and actions (Liu et al. 2019).

[3] $\geq$ 10 million population, as defined by the United Nations. Available at: "World Urbanisation Prospects, The 2018 Revision." UN DESA.

1.2 Urban Green Infrastructure as Nature-Based Solution and Common-Pool Resource

On the one hand, cities in the future will increasingly have to grapple with these multi-dimensional, uncertain, and costly impacts of climate change and environmental degradation (Bai et al. 2018; Dahiya 2012; Gatzweiler et al. 2016). On the other, climate actions led by and in cities are also essential to successful global climate adaptation and environmental restoration, as cities are interconnected, complex adaptive systems that can lead actions through effective governance (Revi et al. 2014).

In addition, urban governments are also uniquely situated where local contexts could be understood, awareness improved and non-state actors mobilised to respond to civil society's pressures (Brunner et al. 2005; Cash and Moser 2000). The advantageous position of cities and the challenges they are facing put urban governments at the centre of successful urban actions, investments, policies, and legal frameworks (Revi et al. 2014).

1.3 Nature-Based Solutions as an Emerging Climate Adaptation Strategy

One of the emerging climate adaptation strategies is *Nature-based Solutions*, also referred to as *Ecosystem-based Adaptations* when it is more specifically referred to in climate adaptation context (Gill et al. 2007a, b; Kabisch et al. 2016). Similarly, Nature-based Solutions and Ecosystem-based Adaptation rely on the installation of urban green infrastructure, including urban green spaces,[4] community parks, street trees, green roofs, green belts, green walls, beehives, urban food garden, ecological corridors, plus the ecosystem services these types of infrastructure provide to mitigate the negative impacts of climate change and deliver positive health benefits (European Environment Agency 2017; Smith and Balakrishnan 2009; World Health Organisation 2017). Urban green infrastructure demonstrates the excellent potential to improve the quality of the urban environment by moderating high temperatures through shading, purifying air, absorbing noise, and filtering and storing rainwater (Pramova et al. 2012). Nature-based Solutions can therefore enhance the resilience of a city facing climate impacts, promote sustainable lifestyles, and improve urban health and well-being (World Health Organisation 2017).

Intergovernmental organisations are actively advocating urban governments to adopt urban green infrastructure as a Nature-based Solution. For instance, UN-Habitat urges local governments to "*apply urban and territorial planning to identify,*

[4] As a component of "green infrastructure," urban green space (Urban green spaces) is defined by the WHO as "all urban land covered by vegetation of any kind. This covers vegetation on private and public grounds, irrespective of size and function, and can also include small water bodies such as ponds, lakes or streams ('blue spaces').".

WHO Europe. 2017. *Urban green spaces: A Brief for Action.* P2.

revitalize, protect and produce high-quality public and green spaces with special ecological or heritage value... and to avoid the creation of heat islands, protect the local biodiversity and support the creation of multifunctional urban green spaces, such as wetlands for rainwater retention and absorption" (UN-Habitat 2015: 22). Awareness on the importance of Nature-based Solutions measures is growing among urban governments (Revi et al. 2014). For example, Portland of Oregon, Philadelphia of Pennsylvania, and Malmö of Sweden, have utilised green roofs, ditches, porous pavements, retention ponds, green space, and downspouts disconnection to reduce stormwater and increase climate resilience, and these methods were much more cost-effective than increasing stormwater storage capacity (Foster et al. 2011; UNFCCC 2015). Many cities in China also have begun to increase urban green spaces under the national scheme "Sponge City" since 2015 under the promotion of its national environmental governance framework *Ecological Civilisation.*

1.4 Urban Green Spaces as a *Common-Pool Resource*

From an economic perspective, urban green spaces such as ecological corridors, parks, and wetlands are *de facto* a type of *common-pool resource (CPR)* for their low *excludability* and high *subtractability* (Colding et al. 2013; Ondrejicka et al. 2017; Ostrom 2005). Urban green spaces are highly subtractable due to the dependence on land availability within an urban territory, which is increasingly scarce in many urban areas. On the other hand, it is challenging to exclude users or residents from enjoying the benefits of urban green spaces, in terms of equity and biological processes. From an equity perspective, health is a fundamental human right as the WHO constitution recognises it.[5] The biological processes determine that all urban residents depend on the ecosystem services provided by urban green spaces for essential health and well-being.

However, at the moment, most academic research on urban green spaces and Nature-based Solutions departs from the perspectives of *urban planning* and *urban ecology*. Insights from urban planning have the potential to improve urban green space accessibility, thereby achieving greater equity, while research from the urban ecology perspective could provide valuable information on ecosystem services and ecological values. Nevertheless, both views focus on the technical dimension of Nature-based Solutions, which cannot address the transaction costs incurred during implementation, such as costs of planning, negotiation, information acquisition, and time.

The concept of *common-pool resources* emerged from the *New Institutional Economics*. This interdisciplinary scientific branch focuses on understanding the complexity of social, political, and commercial life and the changes over time through economics, law, organisation theory, political science, sociology, and anthropology

[5] World Health Organisation. 2017. Human rights and health. Available at: https://www.who.int/news-room/fact-sheets/detail/human-rights-and-health.

(Coase 1998; Klein 1998; Tadelis and Williamson 2012). the New Institutional Economics perspective could provide valuable insights into the institutional aspects of Nature-based Solutions by viewing urban green spaces as a common-pool resource (Colding et al. 2020).

1.5 The Institutional Challenges for Upscaling Nature-Based Solutions in China

Despite increasing evidence of Nature-based Solutions in climate adaptation and urban health improvement and advocacies of intergovernmental organisations, implementing a functional Nature-based Solution sustainable for the long term is not necessarily easier than the so-called hard engineering solutions (Roberts et al. 2012). In addition to the need to reach technical maturity—data collection, expertise, and resource allocation, more complex challenges such as urban land scarcity and institutional lock-ins remain obstacles for many urban governments.

In China, urban land scarcity often hinders planning for new urban green spaces or even just conserving existing ones (Accenture and Chinese Academy of Sciences 2013). Implementing urban green spaces in dense urban areas in China, like in many other countries, has led to a gentrification problem that raised land prices and rent in the neighbour areas (Birge-Liberman 2017; Song and Wu 2010). The primary purpose of urban green spaces is to serve urban development through planning and governance. In China, urban green spaces are treated as a *public good*, a category of urban development land. During the last 30 years, China's urban land resources have been strategically allocated to primarily develop real estate, transport, and commercial infrastructures, which resulted in the country's rapid urbanisation and economic growth.

Currently, many countries or regions govern urban green spaces in combinations of *public goods* (100% accessibility), *club goods* (such as enclosed apartment complexes), and *private goods* (such as private beaches). However, treating urban green spaces as *public goods* overlooks the possibility that their availability could change following the changes in urban land scarcity. Similarly, treating urban green spaces as a *club* or *private good* ignores that urban green spaces benefit all residents' health and well-being. More importantly, governing urban green spaces as a public, club, or private good cannot most adequately reflect the value of its ecosystem services.

Enabling conditions and institutional frameworks for implementing Nature-based Solutions not only need to ensure direct benefits and co-benefits for all residents for greater equity, but also need to reframe the balance between urban population growth, health and well-being, land-driven economic development, and climate resilience (Revi et al. 2014). Such enabling conditions and institutional frameworks entail good urban government, institutional structures, and principles of good governance. Good governance, as argued by many scholars, necessitates institutional values, interests, awareness, analytical and decision-making capacities well fitting into a collaborative working process with other institutions and actors (Birkmann et al. 2010; Bulkeley 2013; Garschagen 2013; Moser and Luers 2007).

Moreover, installing an ecologically functional and well-managed, diverse network of urban green spaces requires a proper understanding of how urban systems and their subsystems function to provide goods and services residents need and desire (Revi et al. 2014). After 30 years of rapid growth, China is currently confronted with prioritising economic development and developing an environmentally more friendly and *people-centred* urbanisation paradigm (Bina 2011; UNDP 2020). The central government has launched each time more ambitious institutional reforms and schemes in the last decade, intending to treat environmental pollution, address climate change impacts, and transition to a more *people-centred* economy (led by a new national development agenda, *Ecological Civilisation*).

However, whether these institutional reforms and schemes can eventually overcome the path dependency or inertia of the economy-centred development strategies practised by Chinese local governments for the last 30 years and achieve real breakthroughs in the challenges mentioned above is a fascinating question for scholars across scientific domains and currently has no definite answers (Garnaut et al. 2018; Yang et al. 2016). The path dependency problem of the current economic model is global, diluting the possibility of achieving critical climate and Sustainable Development Goals.[6] The need for speeding up actions and fundamental shifts is urgent.

1.6 Making the Case of Chinese Cities Seeking NbS as Climate Solutions

Cities stand at the critical forefront of climate challenges as they account for a majority of anthropogenic impacts on Earth and potential solutions, and China has a tremendous responsibility to take bold climate actions to protect its residents and secure its urban infrastructure under its population size and economic trajectory (Bina 2011; Elmqvist et al. 2013a, b; Frantzeskaki et al. 2016). Facing expeditious urbanisation and climate change impacts, how has China governed urban green spaces?

- How do Chinese local governments and other stakeholders interact based on the given rules (mandates, policies) for the planning and management of urban green spaces?
- Are the current institutional settings facilitating Chinese urban governments to plan and manage urban green spaces?

[6] UN News. 16 July 2020. World off track in meeting 2030 Agenda, UN deputy chief warns, calls for solidarity in COVID-19 recovery. Available at: https://news.un.org/en/story/2020/07/1068551.

- How have local governments upscaled urban green spaces towards more effective Nature-based Solutions?
- What is the outcome, regarding the size/quality of urban green spaces or the effectiveness of Nature-based Solutions, and the resultant institutional changes?
- What are the good practices to be shared?
- Why some challenges persisted despite a centralised government system and strong political determination?

1.7 Research Design and Methodology

This book examines China's governance of urban green spaces for effective Nature-based Solutions through an institutional analysis using the Institutional Analysis and Development Framework (IAD Framework) as an analytical framework. The IAD framework was developed by American political scientist and Nobel laureate Elinor Ostrom and her colleagues in the 1980s at the Workshop in Political Theory and Policy Analysis at Indiana University.[7] It is intentionally created to "map" out institutions and help people understand how institutions affect the incentives confronting individuals and their resultant behaviour (Ostrom, 1986, 2005; Kiser and Ostrom, 2000). This framework (see Fig. 3.6: The Institutional Analysis and Development framework contains a broad set of variables that can be used to examine diverse institutional settings in which humans interact, e.g., markets, private enterprises, families, community organisations, legislatures, and government agencies (Ostrom 2005). Researchers throughout the world have applied the IAD Framework to understand the effect of rules on the outcomes of common-pool resource settings. It is compatible with multiple theories in institutional economics, including the transaction cost theory and the common-pool resource theory (Ostrom 2010). Greater explanation can be found in Chap. 4: Research Design and Methodology on how this book analyses (1) **the interaction** between the Chinese local governments and other stakeholders for the planning and management of urban green spaces, and (2) **the outcome of the interaction** through the IAD framework (Fig. 1.1).

This book contains interdisciplinary, qualitative research of three urban green spaces governance case studies through semi-structured, in-depth interviews, archive research, and observation. This research chose three urban green spaces cases to make sense of the changes in urban green spaces governance and the effectiveness of Nature-based Solutions. As mentioned above, the three urban green spaces governance case studies were chosen from Guangzhou, a fast-paced megacity with a 15 million urban population on the low-lying estuary of the Pearl River Delta in Southeast China facing multiple climate change impacts. Facing the current rate of environmental deterioration and climate change impacts, how Guangzhou plans and manages urban green spaces as a Nature-based Solution could provide valuable lessons for itself and many other coastal cities with dense populations. The

[7] Where the PhD Candidate has received a one-year Visiting Scholarship and visited between 2019 and 2020.

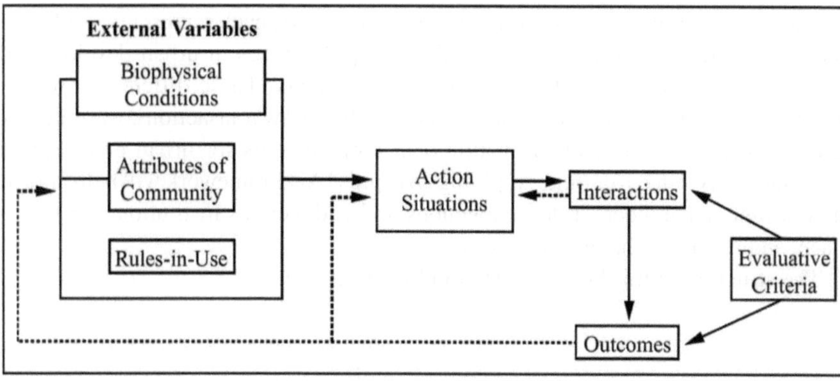

Fig. 1.1 Institutional analysis and development framework. *Source* Adapted from Ostrom (2005: 15)

case studies were chosen within the same city to analyse observed changes in Urban green spaces governance over two decades following the changes of policy mandates within the same government system operated by the same urban government. The researcher interviewed 50 key informants, including academic researchers, urban planners, local government officials, economists, urban consultants, NGOs, and citizens. The interviews were complemented with archive research and review—policy documents, spatial plans, and media reports effective between 2000 and 2018. The three Urban green spaces governance cases in Guangzhou are:

1 **The Panyu Ecological Corridor** (PEC) case presents a conflict of interest between municipal and District Governments. In this case, the Municipal Government of Guangzhou (GZ) aimed to promote a city-wide ecological corridor system; however, the District Government of Panyu (PY) was interested in economic development, which would employ the land nearby or even within the PEC system.
2 **The Haizhu Wetland Park** (HWP) case represents a conflict of interest between GZ, farmers (land and homeowners), and urban migrants (tenants). The land property transfers benefited the farmers but not the urban migrants because of the dual urban–rural household registration system, leaving urban migrants' quality of life unattended.
3 **The Sponge City Demonstration Site (TSCD)** case represents a conflict of interest between different branches within the municipal government, e.g., between the Bureau of Water Affairs and the Administration of Forestry and Gardening. In addition, for the equipment and service suppliers in the private sector with established connections to conventional engineering solutions through public infrastructure construction projects, their vested interests are challenged by the Sponge City Initiative as it is a nature-based rather than an engineering-based solution.

These three cases occurred over the last twenty years, each representing different institutional settings: various combinations of preferences in urban development, property rights bundles, and governance arrangements. These differences will be analysed from an institutional economics perspective: how transaction costs influence actor interactions and resultant institutional arrangements for urban green spaces governance in China over time. Applying Ostrom's IAD Framework will help unravel the relevant stakeholders, rules, exogenous and endogenous institutional factors in place, and resultant changes over time.

The structure of this book is shown in Fig. 1.2:

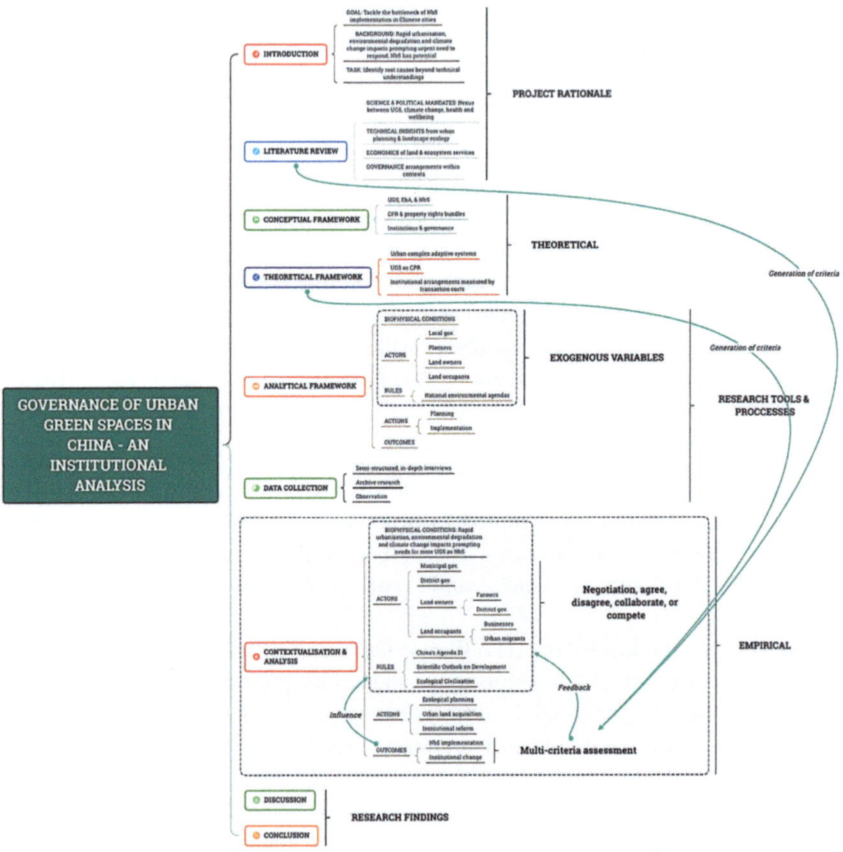

Fig. 1.2 Structure of this book

References

Accenture, & Chinese Academy of Sciences (2013) Towards a new resource economy: promoting the harmonious transformation of Chinese cities (In Chinese: 迈向新资源经济: 推动中国城市和谐转型). https://www.accenture.com/~/media/accenture/conversion-assets/microsites/iframe/new-resource-economy-china/pdf/accenture_report_final.pdf

Angel S, Parent J, Civco DL, Blei A, Potere D (2011) The dimensions of global urban expansion: estimates and projections for all countries, 2000–2050. Prog Plan. https://doi.org/10.1016/j.progress.2011.04.001

Bai X, Chen J, Shi P (2012a) Landscape urbanization and economic growth in China: positive feedbacks and sustainability dilemmas. Environ Sci Technol.https://doi.org/10.1021/es202329f

Bai X, Chen J, Shi P (2012b) Landscape urbanization and economic growth in China: positive feedbacks and sustainability dilemmas. Environ Sci Technolhttps://doi.org/10.1021/es202329f

Bai X, Shi P, Liu Y (2014) Realizing China's urban dream. Naturehttps://doi.org/10.1038/509158a

Bai X, Elmqvist T, Frantzeskaki N, Simon D, Maddox D, Romero-lankao P, Parnell S, Roberts D, Do SW, Know W (2018) In: New integrated urban knowledge for the cities, 462–482

Bina O (2008) Context and systems: thinking more broadly about effectiveness in strategic environmental assessment in China. Environ Managehttps://doi.org/10.1007/s00267-008-9123-5

Bina O (2011) Responsibility for emissions and aspirations for development. In: Garris P (ed) China's responsibility for climate changeEthics, fairness and environmental policy. Policy Press, pp 46–69. https://doi.org/10.1332/policypress/9781847428134.003.0003

Birge-Liberman P (2017) The urban sustainability fix and the rise of the conservancy park. In: Deconstructing the high line: postindustrial urbanism and the rise of the elevated park

Birkmann J, Garschagen M, Kraas F, Quang N (2010) Adaptive urban governance: new challenges for the second generation of urban adaptation strategies to climate change. Sustain Sci. https://doi.org/10.1007/s11625-010-0111-3

Berry HL, Bowen K, Kjellstrom T (2010) Climate change and mental health: a causal pathways framework. Int J Public Health 55(2):123–132. https://doi.org/10.1007/s00038-009-0112-0

Brunner RD, Steelman TA, Coe-Juell L, Cromley CM, Edwards CM, Tucker DW (2005) Adaptive governance: integrating science, policy, and decision making. Local Gov Stud. https://doi.org/10.1017/CBO9781107415324.004

Bulkeley H (2013) Cities and climate change. In: Cities and climate change.https://doi.org/10.4324/9780203077207

Cash DW, Moser SC (2000) Linking global and local scales: designing dynamic assessment and management processes. Glob Environ Chang. https://doi.org/10.1016/S0959-3780(00)00017-0

China's 40 Years of Reform and Development: 1978–2018 (2018) In: Garnaut R, Song L, Fang C (eds) China's 40 years of reform and development: 1978–2018. ANU Press. https://doi.org/10.22459/cyrd.07.2018

Coase R (1998) The new institutional economics. Am Econ Rev 88(2):72–74. http://www.jstor.org/stable/116895

Colding J, Barthel S, Bendt P, Snep R, van der Knaap W, Ernstson H (2013) Urban green commons: Insights on urban common property systems. Glob Environ Chang. https://doi.org/10.1016/j.gloenvcha.2013.05.006

Colding J, Gren Å, Barthel S (2020) The incremental demise of urban green spaces. Land. https://doi.org/10.3390/land9050162

Dahiya B (2012) Cities in Asia, 2012: Demographics, economics, poverty, environment and governance. Cities. https://doi.org/10.1016/j.cities.2012.06.013

Dorning MA, Koch J, Shoemaker DA, Meentemeyer RK (2015) Simulating urbanization scenarios reveals tradeoffs between conservation planning strategies. Landscape Urban Plann.https://doi.org/10.1016/j.landurbplan.2014.11.011

Du Y (2011) Climatic characteristics and change of high temperature in guangdong during the last 49 Years. J Trop Meteorol

Elmqvist T, Goodness J, Marcotullio PJ, Parnell S, Sendstad M, Wilkinson C, Fragkias M, Güneralp B, McDonald RI, Schewenius M, Seto KC (2013a) Urbanization, biodiversity and ecosystem services: challenges and opportunities: a global assessment. In: Urbanization, biodiversity and ecosystem services: challenges and opportunities: a global assessment

Elmqvist T, Goodness J, Marcotullio PJ, Parnell S, Sendstad M, Wilkinson C, Fragkias M, Güneralp B, McDonald RI, Schewenius M, Seto KC (2013b) Urbanization, biodiversity and ecosystem services: challenges and opportunities: a global assessment. In: Urbanization, biodiversity and ecosystem services: challenges and opportunities: a global assessmenthttps://doi.org/10.1007/978-94-007-7088-1

Elsner JB, Kossin JP, Jagger TH (2008) The increasing intensity of the strongest tropical cyclones. Nature 455(7209):92–95. https://doi.org/10.1038/nature07234

European Environment Agency (2017) What is green infrastructure?. https://www.eea.europa.eu/themes/sustainability-transitions/urban-environment/urban-green-infrastructure/what-is-green-infrastructure

Finlay SE, Moffat A, Gazzard R, Baker D, Murray V (2012) Health impacts of wildfires. PLoS Currents. https://doi.org/10.1371/4f959951cce2c

Frantzeskaki N, Kabisch N, McPhearson T (2016) Advancing urban environmental governance: understanding theories, practices and processes shaping urban sustainability and resilience. Environ Sci Policy 62:1–6. https://doi.org/10.1016/j.envsci.2016.05.008

Fritze JC, Blashki GA, Burke S, Wiseman J (2008) Hope, despair and transformation: climate change and the promotion of mental health and wellbeing. Int J Ment Heal Syst. https://doi.org/10.1186/1752-4458-2-13

Foster J, Lowe A, Winkelman S (2011) The value of green infrastructure for urban adaptation. In: The center for clean air policy.https://doi.org/10.1109/MCG.2005.124

Garschagen M (2013) Resilience and organisational institutionalism from a cross-cultural perspective: an exploration based on urban climate change adaptation in Vietnam.https://doi.org/10.1007/s11069-011-9753-4

Gatzweiler FW, Zhu YG, Diez Roux AV, Capon A, Donnelly C, Salem G, Ayad HM, Speizer I, Nath I, Boufford JI, Hanaki K, Rietveld LC, Ritchie P, Jayasinghe S, Parnell S, Zhang Y (n.d.) Advancing health and wellbeing in the changing urban environment (2016th ed.). Zhejiang University Press and Springer

Gill SE, Handley JF, Ennos AR, Pauleit S (2007) Adapting cities for climate change: the role of the green infrastructure. Built Environ 33(1):115–133. https://doi.org/10.2148/benv.33.1.115

Gill SE, Handley JF, Ennos AR, Pauleit S (2007b) Adapting cities for climate change: the role of the green infrastructure. Built Environment. https://doi.org/10.2148/benv.33.1.115

Gong J, Hu Z, Chen W, Liu Y, Wang J (2018) Urban expansion dynamics and modes in metropolitan Guangzhou, China. Land Use Policy 72(October 2016):100–109. https://doi.org/10.1016/j.landusepol.2017.12.025

Haines A, Patz JA (2004) Health effects of climate change. J Am Med Assoc.https://doi.org/10.1001/jama.291.1.99

Hanson S, Nicholls R, Ranger N, Hallegatte S, Corfee-Morlot J, Herweijer C, Chateau J (2011) A global ranking of port cities with high exposure to climate extremes. Clim Change. https://doi.org/10.1007/s10584-010-9977-4

Hallegatte S, Green C, Nicholls RJ, Corfee-Morlot J (2013) Future flood losses in major coastal cities. Nat Clim Chang. https://doi.org/10.1038/nclimate1979

He C, Gao B, Huang Q, Ma Q, Dou Y (2017) Environmental degradation in the urban areas of China: evidence from multi-source remote sensing data. Remote Sens Environ. https://doi.org/10.1016/j.rse.2017.02.027

IPCC (2014) Summary for policymakers. In: Climate change 2014: impacts, adaptation and vulnerability—contributions of the working group II to the fifth assessment report, pp 1–32. https://doi.org/10.1016/j.renene.2009.11.012

Kraas F (2008) Megacities as global risk areas. In: Urban ecology. Springer US, pp 583–596. https://doi.org/10.1007/978-0-387-73412-5_38

Kabisch N, Frantzeskaki N, Pauleit S, Naumann S (2016) Nature-based solutions to climate change mitigation and adaptation in urban areas: perspectives on indicators, knowledge gaps, barriers, and opportunities for action. https://doi.org/10.5751/ES-08373-210239

Kiser LL, Ostrom E (2000) The three worlds of action: a metatheoretical synthesis of institutional approaches. In: Polycentric games and institutions: readings from the workshop in political theory and policy analysis

Klein PG (1998). New Institutional Economics. https://doi.org/10.2139/ssrn.115811

Kuttler W (2008) The urban climate—basic and applied aspects. In: Urban ecology. Springer US, pp 233–248. https://doi.org/10.1007/978-0-387-73412-5_13

Landrigan PJ, Fuller R, Acosta NJR, Adeyi O, Arnold R, Basu N (Nil), Baldé AB, Bertollini R, Bose-O'Reilly S, Boufford JI, Breysse PN, Chiles T, Mahidol C, Coll-Seck AM, Cropper ML, Fobil J, Fuster V, Greenstone M, Haines A, Haines A, Hanrahan D, Hunter D, Khare M, Krupnick A, Lanpher B, Lohani B, Martin K, Mathiasen KV, Mcteer MA, Murray CJL, Ndahimananjara JD, Perera F, Zhong M (2018) The *Lancet* commission on pollution and health. The Lancet 391(10119):462–512. https://doi.org/10.1016/S0140-6736(17)32345-0

Li J, Li C, Zhu F, Song C, Wu J (2013) Spatiotemporal pattern of urbanization in Shanghai, China between 1989 and 2005. Landscape Ecol. https://doi.org/10.1007/s10980-013-9901-1

Liu J, Gatzweiler F, Schmidt L (2019) Addressing the nexus challenge of climate change, human health, and urban sustainability. In: ICS Estudos e Relatórios N°1 / 2019. https://www.ics.uli sboa.pt/flipping/er2019_1/2/

Ma W, Zeng W, Zhou M, Wang L, Rutherford S, Lin H, Liu T, Zhang Y, Xiao J, Zhang Y, Wang X, Gu X, Chu C (2015) The short-term effect of heat waves on mortality and its modifiers in China: an analysis from 66 communities. Environ Int. https://doi.org/10.1016/j.envint.2014.11.004

Malik SM, Awan H, Khan N (2012) Mapping vulnerability to climate change and its repercussions on human health in Pakistan. Glob Health. https://doi.org/10.1186/1744-8603-8-31

Moser SC, Luers AL (2007) Managing climate risks in California: the need to engage resource managers for successful adaptation to change. Clim Change. https://doi.org/10.1007/s10584-007-9384-7

Nicholls RJ, Hanson S, Herweijer C, Patmore N, Hallegatte S, Corfee-Morlot J, Château J, Muir-Wood R (2007) Ranking port cities with high exposure and vulnerability to climate extremes: exposure estimates. In: OECD environment working papers, no 1. https://doi.org/10.1787/011 766488208

Ondrejicka V, Finka M, Husar M, Jamecny L (2017) Urban Space as the commons-new modes for urban space management. IOP Conf Ser: Earth Environ Sci 95:52004. https://doi.org/10.1088/1755-1315/95/5/052004

Ostrom E (1986) An agenda for the study of institutions. Public Choice. https://doi.org/10.1007/BF00239556

Ostrom E (2005) Understanding institutional diversity. Princeton University Press

Ostrom E (2010) Beyond markets and states: polycentric governance of complex economic systems. Am Econ Rev. https://doi.org/10.1257/aer.100.3.641

Pan J (2016) China's environmental governing and ecological civilization. Springer Berlin Heidelberghttps://doi.org/10.1007/978-3-662-47429-7

Pramova E, Locatelli B, Djoudi H, Somorin OA (2012) Forests and trees for social adaptation to climate variability and change. In: Wiley interdisciplinary reviews: climate change. https://doi.org/10.1002/wcc.195

Ren Y, Cui J, Wan S, Liu M, Chen Z, Liao Y, Wang J (2013) climate change impacts on central china and adaptation measures. Adv Clim Change

Rees W, Wackernagel M (2008) Urban Ecological footprints: why cities cannot be sustainable—and why they are a key to sustainability. In: Urban ecology. Springer US, pp 537–555. https://doi.org/10.1007/978-0-387-73412-5_35

Release P (2013) Reducing the €45 billion health cost of air pollution from lorries. Air Qual Clim Change47(2):EI-12–EI-13. http://search.ebscohost.com/login.aspx?direct=true&db=8gh&AN=88314932&site=ehost-live

Revi A, Satterthwaite DE, Aragón-Durand F, Corfee-Morlot J, Kiunsi R, Pelling M, Roberts D, Solecki W (2014) 8. Urban areas. In: Climate change 2014: impacts, adaptation, and vulnerability. part a: global and sectoral aspects. contribution of working group II to the fifth assessment report of the intergovernmental panel on climate change. https://doi.org/10.1017/CBO9781107415379.013

Roberts D, Boon R, Diederichs N, Douwes E, Govender N, Mcinnes A, Mclean C, O'Donoghue S, Spires M (2012) Exploring ecosystem-based adaptation in Durban, South Africa: "learning-by-doing" at the local government coal face. Environ Urban. https://doi.org/10.1177/0956247811431412

State Council of China (2014) National disaster reduction center of the ministry of civil affairs released information about the basic situation of natural disasters nationwide 2013–2017 (in Chinese: 民政部减灾办发布2013年全国自然灾害基本情况)

Schnitzler J, Benzler J, Altmann D, Mücke I, Krause G (2007) Survey on the population's needs and the public health response during floods in Germany 2002. J Public Health Manag Pract. https://doi.org/10.1097/01.PHH.0000285197.23932.3e

Smith, K. R., & Balakrishnan, K. (2009). Mitigating climate, meeting MDGs, and moderating chronic disease: the health co-benefits landscape. *Commonwealth Health Ministers' Update*

Song W, Wu Q (2010) Gentrification and residential differentiation in Nanjing, China. Chinese Geogr Sci. https://doi.org/10.1007/s11769-010-0432-2

Seto KC, Reenberg A, Boone CG, Fragkias M, Haase D, Langanke T, Marcotullio P, Munroe DK, Olah B, Simon D (2012) Urban land teleconnections and sustainability. Proc Natl Acad Sci. https://doi.org/10.1073/pnas.1117622109

Silva RA, West JJ, Lamarque JF, Shindell DT, Collins WJ, Faluvegi G, Folberth GA, Horowitz LW, Nagashima T, Naik V, Rumbold ST, Sudo K, Takemura T, Bergmann D, Cameron-Smith P, Doherty RM, Josse B, MacKenzie IA, Stevenson DS, Zeng G (2017) Future global mortality from changes in air pollution attributable to climate change. Nat Clim Chang 7(9):647–651. https://doi.org/10.1038/nclimate3354

Swim J, Howard G, Clayton S, Reser J, Doherty T, Stern P, Weber E, Gifford R (2017) Psychology & global climate change: addressing a multifaceted phenomenon and set of challenges.https://doi.org/10.4324/9781315314969_2

Tadelis S, Williamson OE (2012) Transaction cost economics. In: The handbook of organizational. https://doi.org/10.4337/9781781001318.00010

UN-Habitat (2015) International Guidelines on Urban and Territorial Planning. https://unhabitat.org/sites/default/files/download-manager-files/IG-UTP_English.pdf

UNFCCC (2015) UNFCCC Adaptation Calendar 2015—Building Climate Resilience in Urban Areas. https://unfccc.int/files/adaptation/application/pdf/calendar_final.pdf

UN DESA (2018) World urbanization prospects: the 2018 revision, key facts. United Nations, Department of Economic and Social Affairs, Pupulation Division. https://doi.org/(ST/ESA/SER.A/366)

United Nations Office for Disaster Risk Reduction (2015) Sendai framework for disaster risk reduction, pp 2015–2030

Walsh J, Wuebbles D, Hayhoe K (2014) Chapter 2: our changing climate. In: Climate change impacts in the United States: the third national climate assessment, October, pp 19–67. https://doi.org/10.7930/J0KW5CXT.On

World Health Organization (2017) Urban green spaces: a brief for action. https://www.euro.who.int/__data/assets/pdf_file/0010/342289/Urban-Green-Spaces_EN_WHO_web3.pdf%3Fua=1

Wu J (2010) Urban sustainability: an inevitable goal of landscape research. Landscape Ecol. https://doi.org/10.1007/s10980-009-9444-7

Wu R, Song X, Chen D, Zhong L, Huang X, Bai Y, Hu W, Ye S, Xu H, Feng B, Wang T, Zhu Y, Fang J, Liu S, Chen J, Wang X, Zhang Y, Huang W (2019) Health benefit of air quality improvement in Guangzhou, China: Results from a long time-series analysis (2006–2016). Environ Int 126:552–559. https://doi.org/10.1016/j.envint.2019.02.064

Xia Y, Guan D, Jiang X, Peng L, Schroeder H, Zhang Q (2016) Assessment of socioeconomic costs to China's air pollution. Atmos Environ. https://doi.org/10.1016/j.atmosenv.2016.05.036

Yang Y, Meng Q, McCarn C, Cooke WH, Rodgers J, Shi K (2016) Effects of path dependencies and lock-ins on urban spatial restructuring in China: a historical perspective on government's role in Lanzhou since 1978. Cities. https://doi.org/10.1016/j.cities.2016.02.011

Yin P, Brauer M, Cohen AJ, Wang H, Li J, Burnett RT, Stanaway JD, Causey K, Larson S, Godwin W, Frostad J, Marks A, Wang L, Zhou M, Murray CJL (2020) The effect of air pollution on deaths, disease burden, and life expectancy across China and its provinces, 1990–2017: an analysis for the Global Burden of Disease Study 2017. The Lancet Planetary Health. https://doi.org/10.1016/s2542-5196(20)30161-3

Chapter 2
Literature Review

2.1 The Theories

2.1.1 Ecological Notion in Western Urban Planning Theories

Urban planning in the post-war period (after 1945) and even before that in Europe and North America was widely perceived as physical planning and design. First, urban planning was primarily concerned with the physical environment instead of the social and economic aspects. Urban planning around that time primarily valued aesthetic design, as architects, civil engineering, and surveying professionals were the main planners. For that reason, the standard of planning then often involved precise "master" plans (Keeble 1969; Taylor 1998) A few examples include the southern extension plan of Amsterdam designed by the early modernist architect Hendrik Petrus Berlage, the *Ville Radieuse* (the Radiant City) by modern architect Le Corbusier, and the Broadacre City urban development concept proposed by American architect Frank Lloyd Wright (Howard 2013). These planning approaches were criticised later for overly emphasising the physical aspects of urban environment and relying on precise and inflexible plans, which resulted in a lack of community consultation and exclusion of social and economic dynamics (Taylor 1998). This period thus was referred to as "utopianism" or "romanticism" (Fishman 2016; Jacobs 1961).

In response, new theories arose in the 1960s to view urban areas as complex systems and urban planning as rational processes (Foley 1973; Taylor 1998). British planning theorist Brian McLoughlin brought forward the "systems view" of planning through his book *Urban and Regional Planning: A Systems Approach* (1969). McLoughlin essentially views and analyses urban areas and regions as systems. McLoughlin essentially views and analyses urban areas and regions as systems. A system, as defined by the Oxford English Dictionary,[1] is "*a set of things working*

[1] https://www.lexico.com/definition/system

© Zhejiang University Press 2023

J. Liu, *Transforming Urban Green Space Governance in China Under Ecological Civilization: An Institutional Analysis*, Urban Health and Wellbeing,
https://doi.org/10.1007/978-981-99-6694-3_2

together as parts of a mechanism or an interconnecting network; a complex whole." McLoughlin argues that components of the urban system include the land's location and human interactions with the land through communications networks. He finds support for systems thinking by reflecting on the interrelatedness of natural phenomena in ecosystems and drawing examples from human-nature interaction failures. Taylor (1998) considers the systems view beneficial for understanding the cities' social and economic dynamics as complex, adaptive systems.

Another emerging urban planning theory throughout the 1960s was the "rational process" view of planning (also commonly known as the "procedural planning theory" into the 1970s), which emphasises the process of planning as a rational process of decision-making. This view is a normative response to the criticism of "blueprint" planning, as it treats planning as a form of social action in which the process and method of planning are the focus, rather than the end-state of plans (Taylor 1998). Compared to the first-generation "physical design" urban planning approach, the complex systems view, and the rational process view of planning demonstrate greater comprehensiveness to both the process and the objects of planning.

However, soon the systems view of urban planning was also criticised for being unnecessarily associated with highly abstract, technical, and mathematical terms such as "modelling" and "optimising," and for treating cities as if they "were some kind of a machine that had one politically uncontentious optimal state" (Taylor 1998: 78). Urban planning critics addressed the need to improve these planning theories by including the social and political actors and their values in context (Friedmann 1969; Scott and Roweis 1977; Thomas 1979).

The progress of urban planning theories in the 1960s and 1970s based on reason and science had its support from the era of modernism, in which planning was viewed as an application of the scientific method to support policymaking (Faludi 1973). The criticisms towards these two views of planning helped reveal the nature of planning as value-laden and political and prompted theorists to shift attention to public participation and value representation of stakeholders in the planning process in Western political and economic context (Arnstein 1969; Long 1959; Pickvance 1977; Taylor 1998). In particular, the critical role of market was examined regarding how it approaches private land property rights and land transactions for planning alongside local planning authorities, and how planning results respond to or serve the need for urban development in liberal economies. Pickvance (1977) argues that market forces are primarily responsible for determining the pattern of land development. But rather than competing with governments, planning theorists generally argue that the role of market was positive and in fact, worked closely with governments to deliver plans which served urban development in liberal economies (Taylor 1998). To conclude, planning theories in the 1970s reached a consensus that urban planning must be viewed, analysed, conducted, and examined rationally through communication with diverse actors within political and economic contexts and that both planning implementation and market forces play crucial roles in determining the outcomes of planning practice (Misgeld et al. 1981).

Entering the 1980s, the "regime theory" emerged to question the assumption of involving government authorities to accomplish desired planning outcomes and explain the increasing cooperation phenomenon between governmental and non-state actors (Stone and Feagin 1990). Regime theory works on the precondition that in liberal societies, often, significant decisions affecting individuals' lives are made by firms operating within the market system rather than the government. Thus, different urban planning regimes may arise in different political and economic circumstances in which degrees of market freedom vary. Another theory—the "problem-centred" planning theory was proposed during the late 1970s to focus more on the issues and problems that urban planning seeks to address. The "problem-centred" planning theory seeks to address the problems of five main areas: (1) the continued economic decline of some urban areas, (2) social divisions and inequalities, (3) the life-threatening ecological changes and planning for greater sustainability, (4) the aesthetic quality of urban environments, and (5) degree of public participation in planning (Taylor 1998).

To conclude, Western urban planning theories evolved (1) from viewing cities as simple physical to complex social-ecological systems, (2) from focusing on the aesthetic and comprehensive "end-state" of planning to the rational process of planning, (3) from political and economic context insensitive to context-sensitive, (4) from local market and civil actors exclusive to local actors inclusive, and (5) from a focus on the method of planning to focus on the practical problems that planning seeks to resolve. During each of these phases, the power of government has been shared further with other social and economic actors. These evolutions of urban planning have gone hand in hand with the changes within the larger Western political and economic contexts, which evolved from post-war reconstruction to modernism further to postmodernism.

2.1.2 Ecological Notion and Governance Philosophy in China

****Taoism is arguably the most acknowledged traditional Chinese philosophy that casts a theoretical influence on spatial planning in China (Gaubatz 1999; Zhang 2008). Tao (道) means "the Way," and it is the primary principle by which all things exist and interrelate. In the Taoist view, each part of the universe is comprised of complementary aspects known as *yin* and *yang*. *Yin* can be described as passive, dark, secretive, negative, weak, feminine, and cool, while *yang*, on the contrary, is active, bright, revealed, positive, masculine, and hot. *Yin* and *yang*, as two contrasting aspects, interact and shift from one extreme to the other constantly, giving rise to the rhythm of nature and lasting change (Ely 2009). The core philosophy of Taoism is primarily embedded in the relation between humans and nature, and "Wu Wei" (无为), which literally means "take no (forceful) action or inexertion," is the central philosophy of Taoism and a proposal for approaching and governing nature. Mountains, for example, have long been imagined as having sacred power, manifesting nature's vital energy, or *qi* (气). Such power gathered the rain clouds that watered farmer's crops and concealed medicinal herbs, fruits, and alchemical minerals, projecting the hope of longevity. Ordinary Chinese people wandered in mountains and practised meditation in caves and grottoes, which were considered gateways to other realms. Caves and grottos are referred to as "cave heavens" (洞天) for spiritual enlightenment and projection of hope for immortality. Poets celebrated the beauty of nature, city dwellers built country estates to escape the dust and pestilence of crowded urban centres, and officials retreated to the mountains as places of refuge during political turmoil. In ancient Chinese thought, nature is conceived as a self-generating, interconnected, complex, and constantly evolving system. This proverb from *Tao De Jing*, the most prominent Taoism classic literature, was written by Laozi (or Lao Tzu): "*Men emulate earth; earth emulates heaven; heaven emulates the Tao; the Tao emulates spontaneity.*" Laozi further advised the governing class (Chapter 64) not to overvalue "*materials that are hard to come by.*" Instead, to desire and learn "*what others do not*" and redeem "*the errors of the masses.*" In the Taoists' view, the essence of Tao is about following the laws of nature and not intervening because the law of nature is the natural intervention that the system needs to maintain a peaceful, natural evolution. So, if a leader could follow the principle of Tao to govern his or her territory and people (Chapter 37), "*everything will develop, evolve, and die by themselves, therefore completing many comprehensive nature cycles.*"

While the Taoism philosophy emphasises human-nature relationships, Confucianism evolved to become the dominant school of thought for the governance of good morals. For instance, in the dialogue between the Duke Jing of Qi and Confucius recorded in Chapter 12:11 of *Confucian Analects*, the Duke asked Confucius about government; Confucius replied: "*Let the ruler be a ruler; the minister, a minister; the father, a father; the son, a son.*" The Duke responded: "*Truly, if the ruler is not a ruler, the subject is not a subject, the father is not a father, and the son is not a son, even if I have revenue, will I enjoy it?*"[2] The Confucian school of thought was essentially a moral education that enabled government officials to exercise proper judgement, as it comprises a complex series of rules and rituals for government officials. The Confucian philosophy functioned well in guiding governance. Officials studied the doctrines to put general rules in specific contexts by their proper or moral judgements (Fukuyama 2016).

In summary, the Taoist philosophy of human-nature harmony and the Confucian school on the moral education of government officials constituted the hallmarks of Chinese central planning and governance (Fukuyama 2016). Under these influences, traditional Chinese urban settlements often had their street networks and primary architecture aligned with the cardinal directions to conform with the Chinese Taoist geomancy and were separated from one another with walls. These traditional Chinese towns or urban areas primarily evolved around clan or family relationships, reflected the complex composition of a hierarchical society, and were described as having applied the classical top-down approach of the "emperor ruling." (Gaubatz 1999).

[2] Chinese Text Project. Available at: https://ctext.org/analects/ens.

Moving onto more modern times, Western urban planning theories and practices have also influenced planning in Chinese cities. The Western rational and comprehensive process view of planning, along with the socialist theory, was adopted in China during the Maoist period between 1949 and the end of the 1970s (Douay 2008; Gaubatz 1999). The year 1949 marked a starting point of a new era for urban planning in China as the new government sought to rebuild a socialist society conformant to Soviet communism. Thus, in the 1950s, China established its planning committees under the influence of socialist theory and Soviet centralism (Gaubatz 1999). In the socialist context, Chinese urban planning delegated state-led urban plans and policies downwards (Wu 2000). In practice, China's urban planning in the 1950s and 1960s was conducted under the notion of "strategic planning," which has an ancient origin in the military domain to help implement strategies in war (Douay 2008). The socialist planning during 1949–late 1970s has been described as centralised socialist planning, or "big government" (Douay 2008). In this period, the central government took major decisions and regulations for urban planning, while municipal officials had limited influence over the form or content of urban development (Zhang 2008). Only a small cohort of traditional actors, namely, government officials and planning experts, were legally responsible for conducting urban planning, and they interacted. This period was thus referred to as urban planning of elitism (Douay 2008; de Jong 2019). Socialist urban planning aimed to "serve the people," but it was often bureaucratic, with planning experts and political elites focusing on technical issues, and was rarely open to external actors (Douay 2008). During this period, the planning principles primarily served political or ideological purposes rather than ecological ones and was mainly operated by the state (Yeh and Wu 1999).

As China shifted to economic liberation reform in 1979, the Western urban regime theory, which explains the increasing influence of private sectors in the urban planning process, also gained popularity in China for understanding the emergence of contemporary cities in the post-Maoist period. Overall, the period of Chinese urban planning and development since the national economic liberation and reform in 1979 has been considered efficiency-driven (Bina 2011). After the economic reform and along the waves of neoliberalism from the 1980s, strategic planning demonstrated even greater influence on Chinese economic development planning, e.g., special economic and technology development zones along China's urban coastline. The Chinese strategic planning in this period also included introducing the market economy system, opening the planning process out to private actors, and reforming land property rights and rules of land transfer. Between the 90s and early 2000, the urban planning realities in China had rapidly evolved to be more complex, thus requiring new theories to understand better the potential of public–private sector collaboration in urban planning (Douay 2008). For this purpose, the urban regime theory proposed by Western urban planning scholars became popular (Douay 2008). During this period, land-dependent and market-led economic development took off and created many new cities, often outpacing the capacity of urban planning land-use regulation institutions (Wang 2015). The mismatch in the pace between market-led development and state-led planning generated multiple social and environmental externalities, e.g., the exclusion of migrant workers from access and the degradation of the urban living environment (Douay 2008). The urban regime theory was popular but, was a lens of political economy rather than ecology. As China enters an era of major governance and technological efforts for urban sustainability, new theories or perspectives are needed to address these problems and guide Chinese urban planning for a socially and environmentally more sustainable way of development (Elmqvist et al. 2013; Wang 2015).

2.2 The Practices

From an urban systems perspective, urban green spaces are viewed as one of the various types of urban infrastructures for public services; hence are referred to as urban green infrastructures (UGI). UGI includes green biological corridors, green roofs, parks, alley, and street trees, and urban forests (bigger parks). UGI are properties of multiple dimensions, including environmental, technical, economic, social, and political/jurisdictional dimensions, which concern a wide range of urban stakeholders—public policymakers, professional planners, private investors, maintenance crews, and users or beneficiaries across all levels and scales (Vincent 2018).

2.2.1 The Environmental Dimension: Nexus Between Urban Green Spaces, Climate Change, Health and Well-Being

Increasing scientific literature supports the nexus between urban green spaces and climate change. urban green spaces can cool down urban areas effectively by shading and reducing evaporation, thereby mitigating the urban heat-island effect and heatwaves and reducing energy consumption (Lindén et al. 2016; Nowak et al. 2017; Oliveira et al. 2011; Zhang et al. 2014, 2017a). If planned and managed adequately, urban green spaces can also provide urban residents with decent access to nature, maintain and protect urban biodiversity, sequestrate CO_2 through photosynthesis, reduce air pollution and noise, retain floodwater and fixate soil, maintain surface moisture, and mitigate the impacts of extreme weather events (see Table 2.1) (Alberti et al. 2018; Escobedo et al. 2010; Hall et al. 2012; Hartig et al. 2014; Strohbach and Haase 2012; Zhang et al. 2017b). Due to these potentials, Urban green spaces are increasingly considered a Nature-based Solution to restore degraded environment, mitigate, and adapt to climate change impacts (Gill et al. 2007a, b; Kabisch et al. 2017).

In addition to mitigating and adapting to climate impacts, the nexus between Urban green spaces and physical and mental health is also evident. Physical inactivity, a likely result of poor walkability and lack of access to recreational areas, is found to be the fourth leading contributor to diseases and accounts for 3.3% of deaths globally (Beaglehole et al. 2011; World Health Organisation, n.d.). Accessible Urban

Table 2.1 Functions of Urban green spaces for climate change mitigation and adaptation

Functions of Urban green spaces	Purpose
Carbon sequestration	Climate mitigation
Energy saving (cooling, biomass)	
Climate regulation	Climate adaptation
Soil fixing	
Stormwater infiltration	
Water and air purification	
Biodiversity conservation	
Cultural services (aesthetic, spiritual, and educational)	

green spaces can facilitate physical activity and relaxation, and form a refuge from noise (González-Oreja et al. 2010; Irvine et al. 2009). Decently distributed trees and vegetation can promote the production of oxygen, and decrease levels of airborne particulate matter and atmospheric CO_2 through photosynthesis and carbon sequestration (Chen 2015; Hartig and Kahn 2016; Lee and Maheswaran 2011; Rowe 2011; Strohbach et al. 2012; Whittinghill et al. 2014). In addition, by promoting physical activity, better air quality and more pleasant temperatures, Urban green spaces can help reduce heart rate, blood pressure, and cortisol levels, hence reducing cardiovascular morbidity and prevalence of type 2 diabetes (Gascon et al. 2016; Lanki et al. 2017; Richardson et al. 2013; Tamosiunas et al. 2014; Ulmer et al. 2016; WHO 2016). Urban green spaces also benefit mental health and cognitive functions, such as improved memory and creativity (Gascon et al. 2015; Nutsford et al. 2013). Hence, making Urban green spaces accessible is a relevant issue, particularly for schoolchildren (Cherrie et al. 2018; Dadvand et al. 2015). Connection with nature significantly affects autonomy, personal growth, and a sense of purpose (Hartig 2008; Hunziker et al. 2007). Furthermore, Urban green spaces can foster social interactions and promote a sense of community, trust, shared norms and values, positive and friendly relationships, and a feeling of belonging (Peters et al. 2010; Ruijsbroek et al. 2017; De Vries et al. 2013). Increasing research has shown the positive association trend between Urban green spaces created and urban health and well-being. However, results have been context-dependent (e.g., coastal, urban, rural), locally focused, and not robust enough to yield efficient policy actions (Kabisch et al. 2017; Scarano 2017).

In summary, increasing scientific literature supports the nexus between Urban green spaces, climate change mitigation and adaptation, and human health and well-being. However, critical insights for policy-making, such as quantification of the economic values, costs, and benefits of Urban green spaces implementation, are existing knowledge gaps yet to be addressed.

2.2.2 *The Technical Dimension*

2.2.2.1 From the Perspectives of Land-Use and Land Economics

The Chinese government has had various approaches towards land-use and urban planning in different phases of economic development in the last seven decades, with the overall tendency of decentralisation and planning based on pragmatic economic valuation. Between 1949 and the end of the 70s, the state had both the ownership and usage right of land during the socialist period; as the government administrated all functions of land, the land value could not be adequately reflected (Lin and Ho 2005).

The economic reform of 1979 resulted in the marketisation of land-use rights to accommodate private sector interests in land-based urban development while the public land ownership was kept, and decentralised decision-making to local urban governments. The marketisation of land-use rights was formalised into what scholars referred to as the asymmetric "dual-track" land market system in the fiscal reforms of the mid-1990s (Lin and Ho 2005). Under this dual-track land system, the economic potential of urban land was more pronounced; cities quickly built many infrastructure facilities conducive to economic development, such as transport, commerce, industry, and housing. The profits brought by the dual-track land system have gradually become an essential source of revenue for local urban governments, as they authorise not only the transactions of land-use rights (or land leasing) but also the conversion of rural land to urban land. These practices have become a major incentive for local governments to approve rural–urban land conversion (Bai et al. 2012b; Ping 2011). Between the 80 s to early Millennium, China's total urban land area increased most significantly; industries and infrastructures promoted by foreign direct investment were the most significant driver for the rural–urban land conversion, also thanks to land subsidies provided by the State government (Deng et al. 2008; Seto and Kaufmann 2003; Tao et al. 2010).

The dual-track land system has broadly contributed to social equity issues, as the benefits of land-based urban economic development remain in cities and are not shared among rural communities (Ding and Lichtenberg 2011). Besides, the large scale of rural land conversion and disordered urbanisation also modified the natural landscape and soil pollution (Chen et al. 2018). Even when local governments began to consider the potential of Urban green spaces for strategic planning in the New Millennium, the primary purposes were to boost land supply and economic productivity and foster more clustered developments to counteract urban sprawl. Scholars have identified social externalities associated with such strategic planning, namely the disconnection of the initial place-based social connections and the displacement of urban migrants (Ng 2019).

2.2.2.2 From the Perspectives of Urban Ecology and Landscape Ecology

From the urban ecology perspective, research on urban systems underwent the stages of examining ecology in, of, and for cities (Childers et al. 2015). Initially, it focused on what is "in" cities, e.g., the terrestrial and aquatic patches within cities and the ecological performances of urban plants and animals (Pickett et al. 2016). More recently, the discipline gained an amplified focus on what is "of" cities, treating entire urban mosaics as social-ecological systems in which biological, social, and built components interact (Pickett et al. 2016). Today, the discipline of urban ecology is recognised as one that seeks to understand urban growth by characterising the biophysical expansion, metabolism, succession, and fluxes of energy and materials in urban areas and urban dwellers. The study of urban ecological systems has attracted much interest, particularly in the last two decades. Scholars revealed limits on the urban-oriented and nature-oriented approaches to understanding the urban environment as a complex, human-dominated system in which urban dwellers play an equal role as the plants and animals in urban territories in influencing urban metabolisms. Mainstream urban ecology focuses on how human activities shape urban environments, including urban climate resulting from different uses of built-up areas, such as the urban-heat-island effect and transport-induced air pollution (Kuttler 2008). As urban populations outnumber rural ones, there is an emergent need also to understand how humans are affected by their urban surroundings (Grimm et al. 2008). In recent literature, urban ecology has been a helpful perspective in revealing the relations between social arrangements (political powers, economic investments, and institutional setups) and the size, distribution, and connectivity of areas which host plants and animals in cities. Studies in urban ecology have revealed significant contributions of natural processes, or ecosystem services, to human health and well-being (Syrbe and Walz 2012). However, questions have persisted regarding the practicality of urban ecology studies for informing policy design, integration, and inspiring civil actions, despite being an interdisciplinary science field (Ahern 2013; Pataki 2015). The questioning, plus recently increasing social-ecological complexities and climate uncertainties in urban environments, prompted the discipline of urban ecology to shift to what scholars referred to as a new paradigm—ecology "for" cities, with greater concern for urban sustainability and a focus on holistic social-ecological systems (Childers et al. 2015; Pickett et al. 2016).

As a similar discipline, landscape ecology was created in the early 1980s to study interactions among landscape elements (Li et al. 2005). Compared to urban ecology, it focuses more specifically on natural ecosystems and offers new concepts, theories, and methods that reveal the importance of spatial patterning and the dynamics of interacting ecosystems (Gill et al. 2007a, b; Turner et al. 2003). The discipline has inspired planners and architects to integrate ecological principles in urban planning and design. Today, landscape ecologists have validated the potential of urban green landscapes, or green infrastructure, in restoring degraded urban environments (Foster et al. 2011; Gill et al. 2007a, b; La Greca et al. 2011; Zhang et al. 2011). Similarly, landscape ecology as a scientific discipline rapidly advanced in the last two decades thanks to emerging computer power and technology such as geographical information systems (GIS) (Turner and Gardner 2015). The emerging need for cities to mitigate and adapt to climate change impacts further underlined the importance of understanding spatial patterns in ecological systems. Landscape ecologists refer to the approach of planning urban natural landscapes and utilising ecosystem services to reduce the vulnerability of ecosystems as "ecosystem-based adaptation" (Ecosystem-based Adaptation) (Revi et al. 2014).

Even though Ecosystem-based Adaptation has been identified as a key contributor to urban resilience, the landscape-oriented or ecology-oriented urban plans have often been criticised as technical and lacking socio-economic stakeholder inclusion (Nesshöver et al. 2017; Revi et al. 2014). Besides, although the Ecosystem-based Adaptation strategy seems to be able to provide multiple ecosystem benefits and is not dependent on high technology, it is far from an easy and affordable approach. The costs of Ecosystem-based Adaptation in urban contexts might be higher than expected largely because costs are higher for land acquisition and ecosystem management (Cartwright et al. 2013; Roberts et al. 2012). For these reasons, scholars call for moving ecology sciences towards an integrated urban science (McPhearson et al. 2016a, b).

2.2.3 The Economic Dimension: Economics of Urban Land and Valuation of Urban Green Spaces

Existing literature has presented several main reasons why governing Urban green spaces remains challenging in China. To begin with, the Chinese urban planning system first commanded and later commodified land primarily as a resource of space (Alix-Garcia et al. 2016). As the central government declared economic reform and liberation in the late 1970s, land-use rights became tradable and quickly commodified for infrastructural development in cities and for building new cities. Rapid land commodification during the country's fast urbanisation process resulted in ambiguous property rights arrangements of urban land in many Chinese cities. In many cases, government officials' delineation of property rights only incurred after the actual occupation or use of urban land (Liu et al. 2019; Zhu 2008). The consequence is that land consumption has occurred at a speed much higher than that of urban population growth. Between 2000 and 2011, the built area in Chinese cities increased by 76.4%, much higher than the urban population growth rate, which increased by 50.5% (State Council of China 2014). In addition, urban land scarcity became increasingly evident, particularly in coastal cities, prompting local officials and planners to emphasise even more on achieving economic efficiency from existing urban land resources (Mao et al. 2020). In general, economists consider land resources the most valuable natural resource considering their scarcity and critical role in producing almost all goods and services crucial for human life (Samuelson and Nordhaus 2010). Land scarcity determines that delivering the highest amount (quantity and quality) of economic goods and services from the land available in any development or management activities is essential.

Furthermore, quantifying the ecological values of Urban green spaces and translating them accurately into monetary terms has been a significant technical difficulty for environmental economists. Costs of technical planning, equipment, and implementation are easily identifiable. However, the costs of potential change to the local ecosystem (increasing or decreasing biodiversity) and potential influence on neighbour property values, residents' health, and well-being cannot be anticipated appropriately (Choumert and Salanié 2008). Current valuation methods of Urban green spaces, e.g., contingent valuation, choice experiment, hedonic price modelling, and the travel cost method, approach the value of Urban green spaces either through proxies or through a hypothetical estimation (Lo and Jim 2010; Powe et al. 2005; Randall et al. 1990). For example, a study on the value of Urban green spaces in Guangzhou, China, focused on residents' willingness to pay for recreation-amenity use (Jim and Chen 2006). While the study showed positive monetary value for the recreation-amenity use of Urban green spaces in Guangzhou, it focused only on the cultural ecosystem services, plus the positive monetary value achieved by the contingent valuation method was hypothetical. Another study focusing on more comprehensive types of ecosystem services in Beijing using the hedonic price modelling method was able to generate suggestions on the ideal patterns and sizes of Urban green spaces for the best ecological outcome (Xu et al. 2016).

Nevertheless, the study lacked compelling cost–benefit reasoning for policymakers. Even when scholars attempt to calculate the potential of Urban green spaces in improving air quality directly, the technical challenge of calculation due to the diversity of, e.g., vegetation configuration, pollutant emissions, meteorology, and tree health, makes it hard to apply the finding directly to policy settings outside the study area (Baró et al. 2014). Hence, these methods are limited in reflecting the total values of urban green spaces. Plus, this type of study still cannot adequately convert the urban green spaces potential into cost–benefit relative terms.

In summary, land resources were first commanded and later commodified for urban development by the Chinese urban planning system, primarily as space resources. Rapid land commodification during the country's fast urbanisation process resulted in ambiguous property rights arrangements of urban land in many Chinese cities. Urban land scarcity became increasingly evident, particularly in coastal cities, prompting local officials and planners to emphasise even more on achieving economic efficiency. Furthermore, quantifying the ecological values of Urban green spaces, which can be helpful for policy-making, has been a significant technical challenge. Finally, implementing urban green spaces to compensate for the negative environmental and social externalities from urban economic development might also lead to externalities, such as gentrification.

2.2.4 *The Social Dimension*

Implementing urban green spaces as a measure to compensate for the negative environmental and social externalities from urban economic development might also lead to externalities of its own. For example, property values can often be increased, intentionally or not, if a decent amount of neighbourhood urban green spaces is provided, resulting in gentrification[3] (Wolch et al. 2014; Cole et al. 2017; Rigolon and Németh 2018). Such urban gentrification often results in various degrees of social exclusion, especially for socio-economically underprivileged people (Gould and Lewis 2012; Seddon et al. 2020; Small et al. 2017). Most existing literature on the social dimension of urban green spaces focuses on Anglo-American settings, and scientific studies on the social impacts of urban green spaces in China only emerged recently (Wolch et al. 2014). China's primary Urban green spaces governance approach has been based on regulatory measures without proper consideration of economic or social costs (Zhu and Zhou 2015). Among the emerging Chinese literature, the social value and performance of Urban green spaces, e.g., accessibility and equity, have been identified as effective indicators of urban green spaces' quality, health, and well-being (Yin et al. 2008). Urban green spaces accessibility has widely accepted measurement metrics, for example, distance to residential areas or the average size of urban green spaces per capita, but intangible concepts such as social values are hard to measure and remain a knowledge gap regardless of contexts.[4]

2.2.5 *The Political Dimension*

2.2.5.1 International Environmental and Climate Governance Frameworks

Numerous international governance frameworks on climate and environment have highlighted the nexus between climate change and urban health, including WHO guidelines, the United Nations Sustainable Development Goals (SDGs), and a few recent accords of the United Nations Framework Convention on Climate Change (UNFCCC). As early as 1986, the WHO launched the Healthy Cities programme, now promoted in thousands of cities worldwide. The programme aimed to set health as an overarching goal on decision-makers agendas and develop comprehensive local

---

[3] According to Investopedia.com, gentrification is "*the transformation of a city neighbourhood from low value to high value. Gentrification is also viewed as a process of urban development in which a neighbourhood or portion of a city develops rapidly in a short period of time, often because of urban renewal programs. This process is often marked by inflated home prices and displacement of a neighbourhood's previous residents.*" Available at: https://www.investopedia.com/terms/g/gentrification.asp#:~:text=Gentrification%20is%20a%20process%20of,costs%20brought%20about%20by%20gentrification.

[4] Cities4Forests: *Gaps in Knowledge*. Available at: https://cities4forests.com/lg-urban-forests/gaps-in-knowledge/

strategies for health protection and sustainable development. The programme functioned well in many cities, with communities gaining a greater understanding of the nexus between health and environment and sectors creating partnerships (Donchin et al. 2006). The 61st World Health Assembly in 2008 led to the creation of the WHO *Work Plan on Climate Change and Health,* which focuses on supporting member states to protect health from climate change with four priorities, including raising awareness, strengthening partnerships, enhancing scientific evidence as well as health systems (WHO 2015). The WHO (2018) frames the *Paris Agreement* launched in 2015 as "potentially the strongest health agreement of the century."[5] The UNFCCC also recognises health as a critical element in its legal framework. The *Paris Agreement* framed the right to health as a fundamental human right in its preamble (United Nations 2015a) and urges member state governments to respect, promote, and consider their respective obligations on the right to health when taking climate actions. Besides, the *Paris Agreement* recognises the central role of mitigation actions and their co-benefits for adaptation, health, and sustainable development, if enhanced responses were taken before 2020 (UN 2015b). Furthermore, it advocates connecting and defining health in the nationally determined contributions (NDCs) for mitigation and adaptation.

The WHO also recognises valuable opportunities for improving health through integrating climate mitigation and adaptation policies across sectors, e.g., energy supply and electricity production, households and buildings, transport, agriculture, and food systems (WHO 2018b). In addition, the WHO identifies potential health gains in infrastructure and governance improvement and recommended a Health in All Policies (HiAP) approach.[6] As for the SDGs, Scholars assert that SDG 11, "Sustainable Cities and Communities" and SDG 3, "Health and Well-being," have a quasi-unidirectional relationship, which means the condition of urban health is highly subjected to the quality of the urban environment. One critical target of SDG 11 is to integrate climate mitigation and adaptation policies substantially into urban development plans by 2020, making SDG 11 synergistic with SDG 13 "Climate Action." Furthermore, SDG 13 can also be synergistic with SDG 3, for a climate-resilient urban development approach could improve environmental quality and thereby support urban health and well-being (Nilsson et al. 2017). The synergy is affirmed by the World Bank Group.[7] Similarly, the Sendai Framework for Disaster Risk Reduction 2015–2030 also underlines the role of urban planning for the implementation of ecosystem-based approaches at both global and regional levels to build resilience.[8]

---

[5] WHO. Climate Change—Overview. Available at: https://www.who.int/health-topics/climate-change#tab=tab_1.

[6] WHO. Health in All Policies—Helsinki Statement Framework for Country Action. Available at: https://apps.who.int/iris/handle/10665/112636.

[7] World Bank Group Climate Change Action Plan. Available at: https://ccacoalition.org/en/resources/world-bank-group-climate-action-plan.

[8] *Sendai Framework for Disaster Risk Reduction 2015–2030* - Priority 2: Strengthening disaster risk governance to manage disaster risk. P18. Available at: https://www.undrr.org/publication/sendai-framework-disaster-risk-reduction-2015-2030.

What is the role of Urban green spaces, then? WHO has primarily emphasised the health potential of Urban green spaces, while the SDGs, as well as the New Urban Agenda (NUA), have primarily accentuated the importance of urban green spaces for equity, and the accords of the UNFCCC have focused on the technical aspects of Urban green spaces as a Nature-based Solution for climate adaptation (UNFCCC 2015). The WHO's review shows the effectiveness of Urban green spaces on human health, and the organisation has actively promoted Urban green spaces as a tool for action for health and well-being (WHO Europe 2016, 2017). The SDGs, on the other hand, have highlighted the role of Urban green spaces, particularly in SDG 3 (health), SDG 11 (cities), and SDG 13 (climate). The potential of Urban green spaces is emphasised primarily in SDG 11: "*Making cities sustainable... involves investment in public transport, creating green public spaces, and improving urban planning and management in participatory and inclusive ways.*"[9] One of the targets of SDG 11 is to provide accessible and inclusive Urban green spaces universally by 2030, particularly for the underprivileged population. Similarly, the NUA recommends cities conduct: "*... participatory, promote civic engagement, engender a sense of belonging and ownership among all their inhabitants, prioritise safe, inclusive, accessible,* ***green and quality public spaces*** *that are friendly for families,..., recognising the specific needs of those in vulnerable situations*" (UN-Habitat 2015: 13; 2016). Furthermore, the UNFCCC has provided technical guidance on planning Urban green spaces for climate adaptation through case studies, tools, and technical reports.[10] Notably, the Global Assessment Report produced by the Intergovernmental Science-Policy Platform on Biodiversity and Ecosystem Services (IPBES) has also endorsed Urban green spaces as a possible pathway to transform urban biodiversity (IPBES 2019).

Together, these international governance frameworks on climate and environment have advised cities to increase urban green spaces as Nature-based Solutions to climate impacts and have shown synergies and interconnections. For instance, SDGs 3, 11, and 13 could play synergistic roles for each other, and urban green spaces could act as a leverage point—adopting urban green spaces as Nature-based Solutions can contribute to mitigating and adapting to climate impacts, as well as improve the urban living environment, thereby bringing health co-benefits. However, the potential of Urban green spaces to serve as a synergistic leverage point to advance climate actions and improve urban health has not been sufficiently highlighted.

2.2.5.2 National Strategies and Action Plans

The nexus between climate change and health also has been well stated in national policies and plans in China, particularly since China surpassed the U.S., became the

[9] Sustainable Development Goals—Goal 11: Make cities inclusive, safe, resilient and sustainable. Available at: https://www.un.org/sustainabledevelopment/cities/

[10] Results from searching "Green Space" on the UNFCCC webpage. Available at: https://www4.unfccc.int/sites/nwpstaging/Pages/Search.aspx?k=green%20space.

biggest CO2 emitter in 2007, and gained international attention on the urban environmental quality from hosting the International Olympic Games in 2008. The central government reframed environmental quality as a public good, claimed responsibility for the provision, and did place a focus on climate action in urban areas. However, the proposition of Urban green spaces as Nature-based Solutions to climate change and health problems is still substantially lacking. The first national action plan to address the environment-health nexus is the *National Environmental Health Action Plan (2007–2015)*, coordinated by the Ministry of Health and 16 other ministerial agencies.[11] The Plan identifies environmental degradation, loss of ecological balance, and subsequent health impacts as key constraints to economic growth and societal development and seeks to establish more people-oriented governance in tackling them. The Plan also proposes action to focus on environmental health impact assessments, early warning systems, and emergency response plans. The first national plan to focus on climate change adaptation is the *National Strategy for Climate Change Adaptation*, which debuted during the Warsaw COP 19 in 2013.[12] It frames human health as one of the seven crucial tasks. It outlines a systematic transition framework focusing on climate change assessment and adaptation planning on national and city levels. The Strategy aims to protect health from increasing risks of many diseases, particularly temperature-sensitive and post-disaster infectious diseases. The same year (2013), the *Action Plan on Prevention and Control of Air Pollution* was approved. It is the first plan to frame decent environmental quality as a public good and as a responsibility of local governments and to set ambitious, specific targets for air pollution reduction. China moved forward with greater determination on climate change actions with the *National Plan on Climate Change (2014–2020)* and the *Action Plan for Urban Adaptation to Climate Change 2014*, both made under "Ecological Civilisation," the country's emerging national environmental governance framework. With the latter, notably, China focused on climate change adaptation in urban areas, concerning the concentrated health and economic impacts.

Similarly, China set the first national medium- to long-term action plan focusing on promoting healthy lifestyles and improving health services and the industry with the *Healthy China 2030 Initiative* launched in 2016.[13] The Plan defines health as an essential condition for socio-economic development and improvement in national health governance as a proactive contribution to the 2030 Sustainable Development Agenda. In actionable terms, *Healthy China 2030* seeks to resonate with the HiAP approach of WHO. *Healthy China 2030* also gained an urban focus with the *Shanghai Consensus on Healthy Cities 2016*, approved by the International Mayors Forum.[14]

In summary, numerous international governance frameworks on climate and environment have highlighted the nexus, including WHO guidelines, the SDGs, and recent accords of the UNFCCC. Together, these frameworks have advised cities to increase urban green spaces as Nature-based Solutions to climate impacts and have

[11] State Council of China (2007).

[12] State Council of China (2013).

[13] WHO (2016b).

[14] WHO (2016c).

shown synergies and interconnections. However, the potential of urban green spaces to serve as a synergistic leverage point to advance climate actions and improve urban health has not been sufficiently highlighted. Similarly, in national policies and plans in China, the nexus between climate change and health also has been well stated. The central government reframed environmental quality as a public good, claimed responsibility for the provision, and did place a focus on climate action in urban areas. However, these action plans still substantially lack the proposition of urban green spaces as Nature-based Solutions to climate change and health problems.

2.3 The Institutional Perspective

2.3.1 Relevance of the Institutional Perspective

Institutional elements and capacities, e.g., the behaviour, values, social norms, and political interests of key actors, authorities, and their legal and regulatory responsibilities, play a critical role in promoting the implementation of urban development and climate adaptation plans (Berkhout 2012; Dietz et al. 2008; Jim 2013; Klein et al. 2015; Mimura et al. 2015; Moser and Ekstrom 2010; Ostrom 2014; Revi et al. 2014; Romero-lankao and Wilhelmi 2018). Studies have documented institutional barriers to planning and implementing climate adaptation, e.g., contesting values, conflicting objectives, tensions, and trade-offs between different policy agendas and priorities (Adger et al. 2009; Storbjörk and Hedrén 2011). In addition to the ignorance of the influence of institutional elements, the scientific challenge to capture institutional dynamics has also been a reason why urban plans often fail when put into practice (Jim 2013; Zhao 2005). In the case of Nature-based Solutions, the lack of capacities and capabilities of local authorities has been identified as a major knowledge gap (Knoblauch et al., n.d.). Complexities of governance networks of multiple actors and institutions are proven barriers to adaptation (Carlsson-Kanyama et al. 2013; Juhola and Westerhoff 2011; Sosa-Rodriguez 2014). The difference in governance approaches, e.g., in how to create adaptation plans, has concrete implications for adaptation outcomes and merits further analysis (Bai et al. 2016, 2018; Mimura et al. 2015).

2.3.2 Critiques on Various Governance Approaches

2.3.2.1 Top-Down and Bottom-Up Governance

The two main tendencies—top-down and bottom-up—have marked the history of urban planning in the past seven decades with an overall shift from top-down towards bottom-up (Healey et al. 1982).

A top-down approach originates from "the top" and is often directed based on decision-making from "the top," while a bottom-up approach describes data collected and processed, decisions made and acted upon at the local level (Nicholls et al. 2015). In urban planning, the top-down governance approach offers advantages of efficiency and economies of scale (Nicholls et al. 2015; Pissourios 2014). Despite so, the appreciation of technical expertise, a characteristic of the top-down approach, has increased (Pissourios 2014). In climate change adaptation, national governments have primarily assumed a coordinating role of adaptation actions for subnational and local-level governments, including information provision, strategies incentivisation, legal and policy frameworks for implementation and supervision, efforts to protect vulnerable groups, and providing financial support for sublevel governments (Mimura et al. 2015). National-level governments have played a key role in climate change adaptation planning and implementation, and subnational governments have played a complementary role to national governments that reflect the governance structure (UK Climate Impacts Programme 2007). The top-down governance approach has received criticism for its tendency to prioritise, overly or sometimes even solely appreciate professional and scientific "expert" knowledge, thereby neglecting local actors and their knowledge and needs and denying citizen participation (Sabatier 1986; Smith 2008).

As a response, the bottom-up governance approach has gained popularity over the years for appreciating and incorporating local people and their knowledge, skills, needs, and experiences into decision-making (Smith 2008). In climate change adaptation planning and implementation, local governments are increasingly recognised as central in addressing the challenges (Blanco et al. 2009; Matthews 2012; Sanchez-Rodriguez 2009). Local authorities can play a significant role in removing barriers to climate adaptation by realigning the incentives of individuals with societal goals, providing the public goods needed for adaptation, or helping with behavioural and cognitive biases (Chambwera et al. 2015). In urban planning, the bottom-up approach is often considered better or more accurate, thanks to the possibility of immediate and small-scale data collection reporting, thereby can be more cost-effective (Nicholls et al. 2015). The bottom-up approach is sometimes called the community-based approach and is endorsed by, e.g., the UNFCCC for its potential in climate change adaptation and resilience-building (IPCC 2014).

Extensive literature has highlighted the need for a comprehensive, institutionally innovative combination of the top-down and bottom-up approaches, as well as a local horizontal network governance approach, to facilitate learning and coordinated actions across multiple authorities (Lowe et al. 2009; Mimura et al. 2015; Pissourios 2014). Despite growing popularity, the bottom-up approach functions well, typically at a small and local scale and has limited capability in coping with the complexities of large scales (Pissourios 2014). Besides, the bottom-up governance approach could fall short for occasions with no "bottom level," for instance, a new settlement or a large city plan expansion with no actual residents yet (Pissourios 2014). In climate change adaptation, local agencies and planners are often confronted by the technical challenge of obtaining credible data on local vulnerabilities and potential impacts and understanding the information correctly (Mimura et al. 2015). For many urban governments, how to cope with the high complexity of climate change impacts across levels within the conventional, often siloed government structures and operational culture across countries remains an operationalisation challenge (Mimura et al. 2015).

2.3.2.2 Multi-level Governance

Compared to the top-down governance approach highlighted by a command-and-comply trait, multi-level governance (MLG) is conducted through more sophisticated rules of power distribution, responsibility sharing, as well as mechanisms and strategies of policy-making (Daniell and Kay 2017). Multi-level governance is an approach with established arrangements for multiple actors to participate in policy-making activities within and across various politico-administrative institutions at different territorial levels (Stephenson 2013). "Multi-level governance" and "cross-level coordination" refer to the same governance approach. One of the earliest and most prominent practices to date in multi-level governance is perhaps the European Integration following the Maastricht Treaty that entered into force in 1992 (Marks 1993), which established the "subsidiarity" principle, particularly in environmental policy, a gesture of political desire to implement policy action at the local level if needed (Golub 1996). Since the emergence of the term, research on multi-level governance has increased over the past three decades, particularly in the governance of non-jurisdictional resources and services that produce cross-border externalities, such as climate change impacts (Quilligan 2008). Similarly, global governance can be considered a form of multi-level governance, for its entirety of political institutions is put forward to de-globalise, solve specific national and regional problems, or provide transnational common goods and services (Zürn 2012).

Scholars consider the multi-level governance approach as having the potential to create positive incentives for sustainable use of resources because of its characteristic of shared responsibility, thereby facilitating learning and adaptation that are context-specific and based on the values and mutual trust of local actors (Adger et al. 2005; Armitage 2007; Newig and Fritsch 2009). A wide range of research, from forest management to urban planning to climate change mitigation and adaptation, widely affirms the need to establish multi-level institutional coordination mechanisms between different political and administrative levels for the effective implementation of policies (Mimura et al. 2015; Revi et al. 2014). Research shows that factors such as the consistency of legal frameworks, power distribution, competencies, and information flows are crucial for increasing trust and empowering local actors for policy integration and implementation (Juhola 2016; Ravikumar et al. 2015). For example, empirical examinations on climate change mitigation and adaptation policy-making processes in the land-use sector in Brazil and Indonesia reveal that cross-level trust building and collaboration between national and subnational levels remain challenging, primarily due to power being disproportionately distributed across governance levels (Di Gregorio et al. 2019). How political power from higher levels impacts the learning and integration of policy-making processes across levels is not yet well understood (Bache 2012; Doherty and Schroeder 2011;

Marquardt 2017). In addition, governments at different levels often have other preferences or priorities: national-level governments engage more extensively in mitigation, and local governments focus more on adaptation (Measham et al. 2011).

2.3.2.3 Collaborative and Participatory Governance

"Collaborative" and "collective" are interchangeable terms. Literature on collaborative governance emerged in recent decades to describe the approach in which public and private stakeholders are brought together for consensus building and planning for environmental management (Ansell and Gash 2007). Contrary to the declining trend of the commanding style in policy processes, collaborative governance relies on face-to-face dialogue, trust building, and the development of commitment and shared understanding to prevail (Ansell and Gash 2007).

Participatory governance seeks to enable citizens to participate in policy processes (Fischer 2012). "Participatory," "democratic," and "inclusive" are interchangeable terms to describe governance approaches that emphasise citizen participation and explore aspects related to citizen competence, empowerment, capacity building, and their impacts on service delivery, equity, and political representation (Ron 2012). Scholars, as well as international organisations, recommend the participatory governance approach for climate change adaptation, as the approach not only does justice to civil rights but also facilitates the incorporation of local resources such as information, knowledge, and finance (Agarwal 2009; Fritz et al. 2019; Tompkins and Eakin 2012; UNDP 2010; World Development Report 2010). For example, many early case studies on forest and water resources governance in Nepal, Japan, India, and Sri Lanka prove that community-participated governance delivers sustainable outcomes (Ostrom 2010). Despite so, governments often have contradictory attitudes towards the participatory governance approach, for citizen participation can inform decision-making, but in the meantime, it increases uncertainty and costs in policy-making and implementation processes (Phillips 2012; Ron 2012).

The two governance approaches share the characteristics of democratic nature and the potential to improve governments' capacity for problem-solving in the long term. In comparison, the collaborative approach tends to demonstrate higher potential in enhancing governments' ability to solve problems as it emphasises more on equal stakeholder collaboration while the participatory governance approach focuses more on the degree of democratic contribution in policy-making processes (Papadopoulos 2012). Research shows that partnerships with the private sector in climate adaptation can increase time and economic efficiency (Engle and Lemos 2010; Tompkins and Eakin 2012). Research on irrigation systems in Nepal and forests worldwide suggests that communication is the key to succeeding in participatory or collaborative governance (Ostrom 2010).

2.3.2.4 The Emerging "Urban" Focus in Governance Literature

Urban governments are responsible for a substantial share of essential infrastructure and service provision. Therefore, their decisions are crucial to the interface between urban economic development and environmental change (Bulkeley and Kern 2006; Hall et al. 2013). The importance of urban governments has been highlighted by increasing literature in the last two decades. Between 1990 and 2015, the literature focused on the "urban" them increased by 14 times roughly (Bai et al. 2018). Scholars seek a greater understanding of the continuous global and local urbanisation processes to resolve emergent, complex environmental and climate change problems (Wolfram et al. 2016). Scholars view cities as crucial actors for addressing climate change impacts and achieving the Paris Agreement (Bai et al. 2018). Often, urban governments face the challenge of balancing the so-called traditional push and pull factors, namely between (1) short-term priorities of economic growth and competitiveness and (2) long-term commitment to restore the urban environment and address climate change impacts (Bai et al. 2014; Leichenko 2011; Romero-Lankao et al. 2017). Besides, scholarly interest in establishing "urban" as a scientific discipline based on complex systems sciences is also rising. This new discipline of urban sciences views cities as self-organised, adaptive, complex socio-ecological or socio-technical systems (McPhearson et al. 2016a, b; Portugali 2006). This discipline considers urbanisation as a multifaceted phenomenon that encompasses profound changes in demographic, land-use, and other resource consumption in environmental, social, cultural, and institutional spheres across local, regional, national, and global scales (Bai et al. 2014, 2018; Friedmann 2006).

2.3.3 Major Institutional Challenges in Urban Governance, Globally and in China

Studies on how institutional dimensions limit or enable policy-making and implementation for climate change adaptation at different levels and in different sectors in both developing and developed countries have increased in recent years (Biesbroek et al. 2013; Engle and Lemos 2010; Glaas and Juhola 2013; Moser and Ekstrom 2010; Rodima-Taylor et al. 2012). Most of the literature has reported on how current institutional arrangements restrict the mainstreaming of climate adaptation across contexts and scales (Biesbroek et al. 2013; Mimura et al. 2015). While middle- and high-income countries report more on institutional barriers such as professional silos, low-income developing countries report on weak institutional environments (Mimura et al. 2015). Several international studies have shown that silos and compartmentalisation between local governments and administrations due to differences in internal norms, values, and priorities are the barriers to mainstreaming climate adaptation policies (van den Berg and Coenen 2012; Wilby and Keenan 2012). For example, studies on water management in multiple municipalities across the world suggest that the current institutional structure and regulatory policies are poorly matched for achieving urban climate adaptation objectives (Aylett 2015; Eisenack and Stecker 2012; Herrfahrdt-Pähle 2013).

In China, rapid economic growth and urban expansion have further highlighted institutional rigidity and mismatch. Studies have identified evidence of responsibility overlaps and policy contradictions due to the silos and compartmentalisation in the administrative system (OECD 2005). When building smart ecocities in China, the environmental and social objectives became harder to achieve as the projects were implemented within the conventional institutional structure. The conventional institutions rely primarily on administrative means of implementation conducted by technical experts and officials and commit more to economic and technological solutions, rather than involving private actors and citizens and focusing on social and environmental well-being (Chai 2015; Xie 2019).

For both developed and developing countries, scholars indicate at least three key areas where institutional arrangements can be improved to facilitate the making and implementation of climate adaptation plans: (1) institutional coordination between different political and administrative levels (2) the horizontal interplay within local governments, and (3) the coordination between formal governmental and administrative agencies and social and private stakeholders (van den Berg and Coenen 2012; Engle and Lemos 2010; Jonsson et al. 2012; Preston et al. 2013; Robinson and Berkes 2011; Runhaar et al. 2012; Wilby and Keenan 2012). While the institutional arrangements in current urban governance across country contexts vary, good governance should facilitate policy and decision processes across public, private, and civil actors and foster the co-production of knowledge, learning, and action over time (Revi et al. 2014; World Bank 2017).

References

Adger WN, Brown K, Tompkins EL (2005) The political economy of cross-scale networks in resource co-management. Ecol Soc. https://doi.org/10.5751/ES-01465-100209

Adger WN, Dessai S, Goulden M, Hulme M, Lorenzoni I, Nelson DR, Naess LO, Wolf J, Wreford A (2009) Are there social limits to adaptation to climate change? Clim Change. https://doi.org/10.1007/s10584-008-9520-z

Agarwal A (2009) Local institutions and adaptation to climate change. In: The social dimensions of climate change: equity and vulnerability in a warming world. https://doi.org/10.1596/978-0-8213-7887-8

Ahern J (2013) Urban landscape sustainability and resilience: the promise and challenges of integrating ecology with urban planning and design. Landsc Ecol. https://doi.org/10.1007/s10980-012-9799-z

Alberti M, McPhearson T, Gonzalez A (2018) Embracing urban complexity. Urban Planet May:45–67. https://doi.org/10.1017/9781316647554.004

Alix-Garcia J, Schneider A, Zhao N (2016) Playing favorites: tax incentives and urban growth in China, 1978–2010. Land Econ. https://doi.org/10.3368/le.92.1.1

Ansell C, Gash A (2007) Collaborative governance in theory and practice. J Publ Admin Res Theory 18(4):543–571. https://doi.org/10.1093/jopart/mum032

Armitage D (2007) Governance and the commons in a multi-level world. Int J Commons. https://doi.org/10.18352/ijc.28
Arnstein SR (1969) A ladder of citizen participation. J Am Plann Assoc. https://doi.org/10.1080/01944366908977225
Aylett A (2015) Institutionalizing the urban governance of climate change adaptation: results of an international survey. Urban Clim. https://doi.org/10.1016/j.uclim.2015.06.005
Bache I (2012) Multi-level governance in the european union. In: Levi-Faur D (ed) The Oxford handbook of governance (Illustrate). Oxford University Press. https://doi.org/10.1093/oxfordhb/9780199560530.013.0044
Bai X, Chen J, Shi P (2012) Landscape urbanization and economic growth in China: positive feedbacks and sustainability dilemmas. Environ Sci Technol. https://doi.org/10.1021/es202329f
Bai X, Shi P, Liu Y (2014) Realizing China's urban dream. Nature. https://doi.org/10.1038/509158a
Bai X, Surveyer A, Elmqvist T, Gatzweiler FW, Güneralp B, Parnell S, Prieur-Richard AH, Shrivastava P, Siri JG, Stafford-Smith M, Toussaint JP, Webb R (2016) Defining and advancing a systems approach for sustainable cities. Curr Opin Environ Sustain. https://doi.org/10.1016/j.cosust.2016.11.010
Bai X, Elmqvist T, Frantzeskaki N, Simon D, Maddox D, Romero-lankao P, Parnell S, Roberts D, Do SW, Know W (2018) In: New integrated urban knowledge for the cities, pp 462–482
Baró F, Chaparro L, Gómez-Baggethun E, Langemeyer J, Nowak DJ, Terradas J (2014) Contribution of ecosystem services to air quality and climate change mitigation policies: the case of urban forests in Barcelona, Spain. Ambio 43(4):466–479. https://doi.org/10.1007/s13280-014-0507-x
Beaglehole R, Bonita R, Horton R, Adams C, Alleyne G, Asaria P, Baugh V, Bekedam H, Billo N, Casswell S, Cecchini M, Colagiuri R, Colagiuri S, Collins T, Ebrahim S, Engelgau M, Galea G, Gaziano T, Geneau R, … Watt J (2011) Priority actions for the non-communicable disease crisis. Lancet. https://doi.org/10.1016/S0140-6736(11)60393-0
Berkhout F (2012) Adaptation to climate change by organizations. Wiley Interdisc Rev: Clim Change. https://doi.org/10.1002/wcc.154
Biesbroek GR, Klostermann JEM, Termeer CJAM, Kabat P (2013) On the nature of barriers to climate change adaptation. Reg Environ Change. https://doi.org/10.1007/s10113-013-0421-y
Bina O (2011) Responsibility for emissions and aspirations for development. In: Garris P (ed) China's responsibility for climate changeEthics, fairness and environmental policy. Policy Press, pp 46–69. https://doi.org/10.1332/policypress/9781847428134.003.0003
Blanco H, Alberti M, Forsyth A, Krizek KJ, Rodríguez DA, Talen E, Ellis C (2009) Hot, congested, crowded and diverse: emerging research agendas in planning. Prog Plan. https://doi.org/10.1016/j.progress.2009.03.001
Bulkeley H, Kern K (2006) Local government and the governing of climate change in Germany and the UK. Urban Stud. https://doi.org/10.1080/00420980600936491
Carlsson-Kanyama A, Carlsen H, Dreborg KH (2013) Barriers in municipal climate change adaptation: results from case studies using backcasting. Futures. https://doi.org/10.1016/j.futures.2013.02.008
Cartwright A, Blignaut J, De Wit M, Goldberg K, Mander M, O'Donoghue S, Roberts D (2013) Economics of climate change adaptation at the local scale under conditions of uncertainty and resource constraints: the case of Durban, South Africa. Environ Urban. https://doi.org/10.1177/0956247813477814
Chai Q (2015) Ecological governance in Chinese megacities (中国超大城市的生态治理). Caixin.Com. http://opinion.caixin.com/2015-01-04/100770467.html?p0#page2
Chambwera M, Heal G, Dubeux C, Hallegatte S, Leclerc L, Markandya A, McCarl BA, Mechler R, Neumann JE, Calvo E, Iglesias A, Navrud S, Kairiza T (2015). Economics of adaptation. In Climate change 2014 impacts, adaptation and vulnerability: Part A: global and sectoral aspects. https://doi.org/10.1017/CBO9781107415379.022
Chen WY (2015) The role of urban green infrastructure in offsetting carbon emissions in 35 major Chinese cities: a nationwide estimate. Cities. https://doi.org/10.1016/j.cities.2015.01.005

Chen J, Guo F, Wang H, Wang Z, Wu Y (2018) Urban land revenue and sustainable urbanization in China: issues and challenges. Sustainability (switzerland). https://doi.org/10.3390/su10072111

Cherrie MPC, Shortt NK, Mitchell RJ, Taylor AM, Redmond P, Thompson CW, Starr JM, Deary IJ, Pearce JR (2018) Green space and cognitive ageing: a retrospective life course analysis in the Lothian Birth Cohort 1936. Soc Sci Med. https://doi.org/10.1016/j.socscimed.2017.10.038

Childers DL, Cadenasso ML, Morgan Grove J, Marshall V, McGrath B, Pickett STA (2015) An ecology for cities: a transformational nexus of design and ecology to advance climate change resilience and urban sustainability. Sustainability (switzerland). https://doi.org/10.3390/su7043774

Choumert J, Salanié J (2008) Provision of urban green spaces: some insights from economics. Landsc Res. https://doi.org/10.1080/01426390802045996

Cole HVS, Lamarca MG, Connolly JJT, Anguelovski I (2017) Are green cities healthy and equitable? Unpacking the relationship between health, green space and gentrification. J Epidemiol Commun Health. https://doi.org/10.1136/jech-2017-209201

Dadvand P, Nieuwenhuijsen MJ, Esnaola M, Forns J, Basagaña X, Alvarez-Pedrerol M, Rivas I, López-Vicente M, De Castro Pascual M, Su J, Jerrett M, Querol X, Sunyer J (2015) Green spaces and cognitive development in primary schoolchildren. Proc Natl Acad Sci. https://doi.org/10.1073/pnas.1503402112

Daniell KA, Kay A (2017) Multi-level governance: an introduction. In Daniell KA, Kay A (eds) Multi-level governance conceptual challenges and case studies from Australia. Australian National University Press

de Jong M (2019) From eco-civilization to city branding: A neo-marxist perspective of sustainable urbanization in China. Sustainability (switzerland). https://doi.org/10.3390/su11205608

De Vries S, van Dillen SME, Groenewegen PP, Spreeuwenberg P (2013) Streetscape greenery and health: stress, social cohesion and physical activity as mediators. Soc Sci Med. https://doi.org/10.1016/j.socscimed.2013.06.030

Deng X, Huang J, Rozelle S, Uchida E (2008) Growth, population and industrialization, and urban land expansion of China. J Urban Econ. https://doi.org/10.1016/j.jue.2006.12.006

Di Gregorio M, Fatorelli L, Paavola J, Locatelli B, Pramova E, Nurrochmat DR, May PH, Brockhaus M, Sari IM, Kusumadewi SD (2019) Multi-level governance and power in climate change policy networks. Glob Environ Chang 54:64–77. https://doi.org/10.1016/J.GLOENVCHA.2018.10.003

Dietz T, Ostrom E, Stern PC (2008) The struggle to govern the commons. In: Urban ecology: an international perspective on the interaction between humans and nature. https://doi.org/10.1007/978-0-387-73412-5_40

Ding C, Lichtenberg E (2011) Land and urban economic growth in China. J Reg Sci. https://doi.org/10.1111/j.1467-9787.2010.00686.x

Doherty E, Schroeder H (2011) Forest tenure and multi-level governance in avoiding deforestation under REDD+. Glob Environ Polit 11(4):66–88. https://doi.org/10.1162/GLEP_a_00084

Donchin M, Shemesh AA, Horowitz P, Daoud N (2006) Implementation of the healthy cities' principles and strategies: an evaluation of the Israel healthy cities network. Health Promot Int. https://doi.org/10.1093/heapro/dal024

Douay N (2008) Shanghai: urban planning styles in evolution. China Perspect 2008(4):16–25. https://login.e.bibl.liu.se/login?url=https://search.ebscohost.com/login.aspx?direct=true&db=aph&AN=39342063&site=eds-live&scope=site

Eisenack K, Stecker R (2012) A framework for analyzing climate change adaptations as actions. Mitig Adapt Strat Glob Change. https://doi.org/10.1007/s11027-011-9323-9

Ely B (2009) Change and continuity: the influences of taoist philosophy and cultural practices on contemporary art practice. University of Western Sydney. https://researchdirect.westernsydney.edu.au/islandora/object/uws%3A6834/datastream/PDF/view

Engle NL, Lemos MC (2010) Unpacking governance: Building adaptive capacity to climate change of river basins in Brazil. Glob Environ Chang. https://doi.org/10.1016/j.gloenvcha.2009.07.001

Escobedo F, Varela S, Zhao M, Wagner JE, Zipperer W (2010) Analyzing the efficacy of subtropical urban forests in offsetting carbon emissions from cities. Environ Sci Policy. https://doi.org/10.1016/j.envsci.2010.03.009

Faludi A (1973) A reader in planning theory—A volume in Urban and regional planning series. Pergamon Press. https://doi.org/10.1016/C2009-0-10937-6

Fischer F (2012) Participatory governance: from theory to practice. In Levi-Faur D (ed) The Oxford handbook of governance (illustrate). Oxford University Press. https://doi.org/10.1093/oxfordhb/9780199560530.013.0032

Fishman R (2016) Urban utopias in the twentieth century. Howard E, Wright FL, Corbusier L (eds) In: Readings in planning theory, 4h edn. https://doi.org/10.1002/9781119084679.ch1

Foley DL (1973) British town planning: one ideology or three? In: A reader in planning theory. https://doi.org/10.1016/B978-0-08-017066-4.50010-2

Foster J, Lowe A, Winkelman S (2011) The value of green infrastructure for urban adaptation. In: The Center for Clean Air Policy. https://doi.org/10.1109/MCG.2005.124

Friedmann J (1969) Notes on societal action. J Am Inst Plann 35(5):311–318. https://doi.org/10.1080/01944366908977241

Friedmann J (2006) Four theses in the study of China's urbanization. Int J Urban Reg Res. https://doi.org/10.1111/j.1468-2427.2006.00671.x

Fritz S, See L, Carlson T, Haklay M (Muki), Oliver JL, Fraisl D, Mondardini R, Brocklehurst M, Shanley LA, Schade S, Wehn U, Abrate T, Anstee J, Arnold S, Billot M, Campbell J, Espey J, Gold M, Hager G, … West S (2019) Citizen science and the united nations sustainable development goals. Nat Sustain. https://doi.org/10.1038/s41893-019-0390-3

Fukuyama F (2016) Reflections on Chinese governance, J Chin Gov 1(3):379–391. https://doi.org/10.1080/23812346.2016.1212522

Gar-On Yeh A, Wu F (1999) The transformation of the urban planning system in China from a centrally-planned to transitional economy. Prog Plan 51(3):167–252. https://doi.org/10.1016/S0305-9006(98)00029-4

Gascon M, Mas MT, Martínez D, Dadvand P, Forns J, Plasència A, Nieuwenhuijsen MJ (2015) Mental health benefits of long-term exposure to residential green and blue spaces: a systematic review. Int J Environ Res Public Health. https://doi.org/10.3390/ijerph120404354

Gascon M, Triguero-Mas M, Martínez D, Dadvand P, Rojas-Rueda D, Plasència A, Nieuwenhuijsen MJ (2016) Residential green spaces and mortality: a systematic review. Environ Int. https://doi.org/10.1016/j.envint.2015.10.013

Gaubatz P (1999) China's urban transformation: patterns and processes of morphological change in Beijing, Shanghai and Guangzhou. Urban Stud 36(9):1495–1521. https://doi.org/10.1080/0042098992890

Gill SE, Handley JF, Ennos AR, Pauleit S (2007b) Adapting cities for climate change: The role of the green infrastructure. Built Environ. https://doi.org/10.2148/benv.33.1.115

Gill SE, Handley JF, Ennos AR, Pauleit S (2007) Adapting cities for climate change: the role of the green infrastructure. Built Environ 33(1):115–133. https://doi.org/10.2148/benv.33.1.115

Glaas E, Juhola S (2013) New levels of climate adaptation policy: analyzing the institutional interplay in the Baltic sea region. Sustainability (switzerland). https://doi.org/10.3390/su5010256

Golub J (1996) Sovereignty and subsidiarity in EU environmental policy. Polit Stud 44(4):686–703. https://doi.org/10.1111/j.1467-9248.1996.tb01749.x

González-Oreja JA, Bonache-Regidor C, De La Fuente-Díaz-Ordaz AA (2010) Far from the noisy world? Modelling the relationships between park size, tree cover and noise levels in urban green spaces of the city of Puebla, Mexico. Interciencia. https://doi.org/0378-1844/10/07/486-07

Gould K, Lewis T (2012) The environmental injustice of green gentrification. … , Immigration, and ethnic politics in a ….

Hall JW, Henriques JJ, Hickford AJ, Nicholls RJ (2013) Systems-of-systems analysis of national infrastructure. Proc Inst Civil Eng Eng Sustain. https://doi.org/10.1680/ensu.12.00028

Hall JM, Handley JF, Ennos AR (2012) The potential of tree planting to climate-proof high density residential areas in Manchester, UK. Landsc Urban Plan. https://doi.org/10.1016/j.landurbplan.2011.11.015

Hartig T (2008) Green space, psychological restoration, and health inequality. Lancet. https://doi.org/10.1016/S0140-6736(08)61669-4

Hartig T, Kahn PH (2016) Living in cities, naturally. Science. https://doi.org/10.1126/science.aaf3759

Hartig T, Mitchell R, de Vries S, Frumkin H (2014) Nature and Health. Annu Rev Public Health 35(1):207–228. https://doi.org/10.1146/annurev-publhealth-032013-182443

Healey P, McDougall G, Thomas MJ (1982) Theoretical DEcosystem-based adaptationtes in planning: towards a coherent dialogue. In: Planning theory prospects for the 1980s, 1st edn. Pergamon Press, pp 5–22

Herrfahrdt-Pähle E (2013) Integrated and adaptive governance of water resources: the case of South Africa. Reg Environ Change. https://doi.org/10.1007/s10113-012-0322-5

Howard E (2013) Garden cities of to-morrow. In: Garden cities of to-morrow. https://doi.org/10.4324/9780203716779

Hunziker M, Buchecker M, Hartig T (2007) Space and place—Two aspects of the human-landscape relationship. Challenges Landsc Res. https://doi.org/10.1007/978-1-4020-4436-6_5

Intergovernmental Science-Policy Platform on Biodiversity and Ecosystem Services (2019) Global assessment report on biodiversity and ecosystem service. In: DEcosystem-based adaptationting nature's value. https://doi.org/10.1007/978-3-319-99244-0_2

IPCC (2014) Summary for policymakers. climate change 2014: impacts, adaptation and vulnerability—Contributions of the working group II to the fifth assessment report, pp 1–32. https://doi.org/10.1016/j.renene.2009.11.012

Irvine KN, Devine-Wright P, Payne SR, Fuller RA, Painter B, Gaston KJ (2009) Green space, soundscape and urban sustainability: an interdisciplinary, empirical study. Local Environ. https://doi.org/10.1080/13549830802522061

Jacobs J (1961) The death and life of Great American cities (1992 repri). Vintage Books

Jim CY (2013) Sustainable urban greening strategies for compact cities in developing and developed economies. Urban Ecosyst 16(4):741–761. https://doi.org/10.1007/s11252-012-0268-x

Jim CY, Chen WY (2006) Recreation-amenity use and contingent valuation of urban greenspaces in Guangzhou, China. Landsc Urban Plan. https://doi.org/10.1016/j.landurbplan.2004.08.008

Jonsson AC, Hjerpe M, Andersson-Sköld Y, Glaas E, André K, Simonsson L (2012) Cities' capacity to manage climate vulnerability: Experiences from participatory vulnerability assessments in the lower Göta Älv Catchment. Local Envir, Sweden. https://doi.org/10.1080/13549839.2012.685880

Juhola S (2016) Barriers to the implementation of climate change adaptation in land use planning. Int J Clim Change Strateg Manage 8(3):338–355. https://doi.org/10.1108/IJCCSM-03-2014-0030

Juhola S, Westerhoff L (2011) Challenges of adaptation to climate change across multiple scales: a case study of network governance in two European countries. Environ Sci Policy. https://doi.org/10.1016/j.envsci.2010.12.006

Kabisch N, van den Bosch M, Lafortezza R (2017) The health benefits of nature-based solutions to urbanization challenges for children and the elderly—A systematic review. Environ Res. https://doi.org/10.1016/j.envres.2017.08.004

Keeble LB (1969) Principles and practice of town and country planning, 4th repri. Estates Gazette. https://books.google.com/books/about/Principles_and_practice_of_town_and_coun.html?id=7i0RAQAAMAAJ

Klein RJT, Midgley GF, Preston BL, Alam M, Berkhout FGH, Dow K, Shaw MR, Gitay H, Thurlow J, Buob S, Thomas A (2015) Adaptation opportunities, constraints, and limits. In: Climate change 2014 impacts, adaptation and vulnerability: Part A: global and sectoral aspects. https://doi.org/10.1017/CBO9781107415379.021

Kuttler W (2008) The urban climate—Basic and applied aspects. In: Urban ecology. Springer, US, pp 233–248. https://doi.org/10.1007/978-0-387-73412-5_13

La Greca P, La Rosa D, Martinico F, Privitera R (2011) Agricultural and green infrastructures: the role of non-urbanised areas for eco-sustainable planning in a metropolitan region. Environ Pollut. https://doi.org/10.1016/j.envpol.2010.11.017

Lanki T, Siponen T, Ojala A, Korpela K, Pennanen A, Tiittanen P, Tsunetsugu Y, Kagawa T, Tyrväinen L (2017) Acute effects of visits to urban green environments on cardiovascular physiology in women: a field experiment. Environ Res. https://doi.org/10.1016/j.envres.2017.07.039

Lee ACK, Maheswaran R (2011) The health benefits of urban green spaces: a review of the evidence. J Public Health. https://doi.org/10.1093/pubmed/fdq068

Leichenko R (2011) Climate change and urban resilience. Cur Opin Environ Sustain. https://doi.org/10.1016/j.cosust.2010.12.014

Li F, Wang R, Paulussen J, Liu X (2005) Comprehensive concept planning of urban greening based on\necological principles: a case study in Beijing, China. Landsc Urban Plan. https://doi.org/10.1016/j.landurbplan.2004.04.002

Lin GCS, Ho SPS (2005) The state, land system, and land development processes in contemporary China. Ann Assoc Am Geogr. https://doi.org/10.1111/j.1467-8306.2005.00467.x

Lindén J, Fonti P, Esper J (2016) Temporal variations in microclimate cooling induced by urban trees in Mainz. Urban Forestry and Urban Greening, Germany. https://doi.org/10.1016/j.ufug.2016.09.001

Liu J, Gatzweiler F, Schmidt L (2019) Addressing the nexus challenge of climate change, human health, and urban sustainability. ICS Estudos e Relatórios Nº1/2019. https://www.ics.ulisboa.pt/flipping/er2019_1/2/

Lo AY, Jim CY (2010) Willingness of residents to pay and motives for conservation of urban green spaces in the compact city of Hong Kong. Urban Forest Urban Green. https://doi.org/10.1016/j.ufug.2010.01.001

Long NE (1959) Planning and politics in urban development. J Am Inst Plann 25(4):167–169. https://doi.org/10.1080/01944365908978328

Lowe A, Foster J, Winkelman S (2009) Ask the climate question: adapting to climate change impacts in urban regions. Environ Protect

Mao X, Huang X, Song Y, Zhu Y, Tan Q (2020) Response to urban land scarcity in growing megacities: urban containment or inter-city connection? Cities. https://doi.org/10.1016/j.cities.2019.102399

Marks G (1993) Structural policy and multilevel governance in the EC. In: The state of the european community. Vol. 2: the Maastricht DEcosystem-based adaptationtes and beyond

Marquardt J (2017) Conceptualizing power in multi-level climate governance. J Clean Prod 154:167–175. https://doi.org/10.1016/J.JCLEPRO.2017.03.176

Matthews T (2012) Responding to climate change as a transformative stressor through metro-regional planning. Local Environ. https://doi.org/10.1080/13549839.2012.714764

McLoughlin J (1969) Urban and regional planning. a systems approach. In: Urban and regional planning. a systems approach ….. https://doi.org/10.1017/CBO9781107415324.004

McPhearson T, Haase D, Kabisch N, Gren Å (2016a) Advancing understanding of the complex nature of urban systems. Ecol Indicators. https://doi.org/10.1016/j.ecolind.2016.03.054

McPhearson T, Pickett STA, Grimm NB, Niemelä J, Alberti M, Elmqvist T, Weber C, Haase D, Breuste J, Qureshi S (2016b) Advancing urban ecology toward a science of cities. Bioscience. https://doi.org/10.1093/biosci/biw002

Measham TG, Preston BL, Smith TF, Brooke C, Gorddard R, Withycombe G, Morrison C (2011) Adapting to climate change through local municipal planning: barriers and challenges. Mitig Adapt Strat Glob Change. https://doi.org/10.1007/s11027-011-9301-2

Mimura N, Pulwarty RS, Duc DM, Elshinnawy I, Redsteer MH, Huang HQ, Nkem JN, Rodriguez RAS, Moss R, Vergara W, Darby LS, Kato S (2015) Adaptation planning and implementation.

In: Climate change 2014 impacts, adaptation and vulnerability: Part A: global and sectoral aspects. https://doi.org/10.1017/CBO9781107415379.020
Misgeld D, Habermas J, McCarthy T (1981) Communication and the evolution of society. Contemp Sociol. https://doi.org/10.2307/2067761
Moser SC, Ekstrom JA (2010) A framework to diagnose barriers to climate change adaptation. Proc Natl Acad Sci. https://doi.org/10.1073/pnas.1007887107
Nesshöver C, Assmuth T, Irvine KN, Rusch GM, Waylen KA, Delbaere B, Haase D, Jones-Walters L, Keune H, Kovacs E, Krauze K, Külvik M, Rey F, van Dijk J, Vistad OI, Wilkinson ME, Wittmer H (2017) The science, policy and practice of nature-based solutions: an interdisciplinary perspective. Sci Total Environ. https://doi.org/10.1016/j.scitotenv.2016.11.106
Newig J, Fritsch O (2009) Environmental governance: participatory, multi-level—And effective? Environ Policy Gov. https://doi.org/10.1002/eet.509
Ng MK (2019) Governing green urbanism: the case of Shenzhen, China. J Urban Affairs https://doi.org/10.1080/07352166.2016.1271623
Nicholls D, Barnes F, Acrea F, Chen C, Buluç LY, Parker MM (2015). General technical report PNW-GTR-906: top-down and bottom-up approaches to greenhouse gas inventory methods a comparison between national-and forest-scale reporting methods. https://www.fs.fed.us/pnw/pubs/pnw_gtr906.pdf
Nilsson M, McCollum D, Echeverri LG, Riahi K, Parkinson S (2017) A guide to SDG interactions : from science to implementation. In: A guide To SDG interactions : from science to implementation. https://doi.org/10.24948/2017.01
Nowak DJ, Appleton N, Ellis A, Greenfield E (2017) Residential building energy conservation and avoided power plant emissions by urban and community trees in the United States. Urban Forest Urban Green. https://doi.org/10.1016/j.ufug.2016.12.004
Nutsford D, Pearson AL, Kingham S (2013) An ecological study investigating the association between access to urban green space and mental health. Public Health. https://doi.org/10.1016/j.puhe.2013.08.016
OECD (2005) China in the global economy: governance in China
Oliveira S, Andrade H, Vaz T (2011) The cooling effect of green spaces as a contribution to the mitigation of urban heat: a case study in Lisbon. Build Environ. https://doi.org/10.1016/j.buildenv.2011.04.034
Ostrom E (2010) Beyond markets and states: Polycentric governance of complex economic systems. Am Econ Rev. https://doi.org/10.1257/aer.100.3.641
Ostrom E (2014) A polycentric approach for coping with climate change. Ann Econ Financ. https://doi.org/10.2139/ssrn.1934353
Papadopoulos Y (2012) The democratic quality of collaborative governance. In: Levi-Faur D (ed) The Oxford handbook of governance (Illustrate). Oxford University Press. https://doi.org/10.1093/oxfordhb/9780199560530.013.0036
Pataki DE (2015) Grand challenges in urban ecology. Front Ecol Evol. https://doi.org/10.3389/fevo.2015.00057
Peters K, Elands B, Buijs A (2010) Social interactions in urban parks: stimulating social cohesion? Urban Fores Urban Green. https://doi.org/10.1016/j.ufug.2009.11.003
Phillips SD (2012) The new citizenship and governance: alternative intersections. In: Levi-Faur D (ed) The Oxford handbook of governance (Illustrate). Oxford University Press. https://doi.org/10.1093/oxfordhb/9780199560530.013.0034
Pickett STA, Cadenasso ML, Childers DL, Mcdonnell MJ, Zhou W (2016) Evolution and future of urban ecological science: ecology in, of, and for the city. Ecosyst Health Sustain. https://doi.org/10.1002/ehs2.1229
Pickvance CG (1977) Physical planning and market forces in urban development. Natl Westminster Bank Quart Rev 41–50
Ping YC (2011) Explaining land use change in a Guangdong County: the supply side of the story. China Q. https://doi.org/10.1017/S0305741011000683

Pissourios IA (2014) Top-down and bottom-up urban and regional planning: towards a framework for the use of planning standards. Eur Spatial Res Policy. https://doi.org/10.2478/esrp-2014-0007

Portugali J (2006) Complexity theory as a link between space and place. Environ Plan A. https://doi.org/10.1068/a37260

Powe NA, Garrod GD, McMahon PL (2005) Mixing methods within stated preference environmental valuation: Choice experiments and post-questionnaire qualitative analysis. Ecol Econ. https://doi.org/10.1016/j.ecolecon.2004.06.022

Preston BL, Dow K, Berkhout F (2013) The climate adaptation frontier. Sustainability (switzerland). https://doi.org/10.3390/su5031011

Programme UCI (2007) Identifying adaptation options. Policy Paper. https://doi.org/10.3389/fphar.2018.00213

Quilligan JB (2008) Global commons goods. Civil society as global commons organizations. KOSMOS J. https://www.kosmosjournal.org/article/global-commons-goods-civil-society-as-global-commons-organizations/

Randall A, Mitchell RC, Carson RT (1990) Using surveys to value public goods: the contingent valuation method. Land Econ. https://doi.org/10.2307/3146688

Ravikumar A, Larson AM, Duchelle AE, Myers R, Tovar JG (2015) Multilevel governance challenges in transitioning towards a national approach for REDD+: evidence from 23 subnational REDD+ initiatives. Int J Commons. https://doi.org/10.18352/ijc.593

Revi A, Satterthwaite DE, Aragón-Durand F, Corfee-Morlot J, Kiunsi R, Pelling M, Roberts D, Solecki W (2014) 8. Urban areas. In: Climate change 2014: impacts, adaptation, and vulnerability. Part A: global and sectoral aspects. contribution of working group II to the fifth assessment report of the intergovernmental panel on climate change. https://doi.org/10.1017/CBO9781107415379.013

Richardson EA, Pearce J, Mitchell R, Kingham S (2013) Role of physical activity in the relationship between urban green space and health. Public Health. https://doi.org/10.1016/j.puhe.2013.01.004

Rigolon A, Németh J (2018) "We're not in the business of housing:" Environmental gentrification and the nonprofitization of green infrastructure projects. Cities. https://doi.org/10.1016/j.cities.2018.03.016

Roberts D, Boon R, Diederichs N, Douwes E, Govender N, Mcinnes A, Mclean C, O'Donoghue S, Spires M (2012) Exploring ecosystem-based adaptation in Durban, South Africa: "learning-by-doing" at the local government coal face. Environ Urban. https://doi.org/10.1177/0956247811431412

Robinson LW, Berkes F (2011) Multi-level participation for building adaptive capacity: Formal agency-community interactions in northern Kenya. Glob Environ Chang. https://doi.org/10.1016/j.gloenvcha.2011.07.012

Rodima-Taylor D, Olwig MF, Chhetri N (2012) Adaptation as innovation, innovation as adaptation: an institutional approach to climate change. Appl Geogr. https://doi.org/10.1016/j.apgeog.2011.10.011

Romero-Lankao P, McPhearson T, Davidson DJ (2017) The food-energy-water nexus and urban complexity. Nat Clim Chang. https://doi.org/10.1038/nclimate3260

Romero-lankao P, Wilhelmi O (2018) Chapter 4 : Live with risk while reducing vulnerability. May:92–112

Ron A (2012) Modes of democratic governance. In: Levi-Faur D (ed) The Oxford handbook of governance (Illustrate). Oxford University Press. https://doi.org/10.1093/oxfordhb/9780199560530.013.0033

Rowe DB (2011) Green roofs as a means of pollution abatement. Environ Pollut. https://doi.org/10.1016/j.envpol.2010.10.029

Ruijsbroek A, Mohnen SM, Droomers M, Kruize H, Gidlow C, Gražulevičiene R, Andrusaityte S, Maas J, Nieuwenhuijsen MJ, Triguero-Mas M, Masterson D, Ellis N, van Kempen E, Hardyns W, Stronks K, Groenewegen PP (2017) Neighbourhood green space, social environment and

mental health: an examination in four European cities. Int J Public Health. https://doi.org/10.1007/s00038-017-0963-8
Runhaar H, Mees H, Wardekker A, van der Sluijs J, Driessen PPJ (2012) Adaptation to climate change-related risks in Dutch urban areas: stimuli and barriers. Reg Environ Change. https://doi.org/10.1007/s10113-012-0292-7
Sabatier PA (1986) Top-down and bottom-up approaches to implementation research: a critical analysis and suggested synthesis. J Publ Policy. https://doi.org/10.1017/S0143814X00003846
Samuelson PA, Nordhaus WD (2010) Microeconomics, 19th edn. McGraw-Hill/Irwin
Sanchez-Rodriguez R (2009) Learning to adapt to climate change in urban areas. A review of recent contributions. Curr Opin Environ Sustain. https://doi.org/10.1016/j.cosust.2009.10.005
Scarano FR (2017) Ecosystem-based adaptation to climate change: concept, scalability and a role for conservation science. Perspect Ecol Conserv. https://doi.org/10.1016/j.pecon.2017.05.003
Scott AJ, Roweis ST (1977) Urban planning in theory and practice: a reappraisal. Environ Plan A. https://doi.org/10.1068/a091097
Seddon N, Chausson A, Berry P, Girardin CAJ, Smith A, Turner B (2020) Understanding the value and limits of nature-based solutions to climate change and other global challenges. Philos Trans Roy Soc B: Biol Sci. https://doi.org/10.1098/rstb.2019.0120
Seto KC, Kaufmann RK (2003) Modeling the drivers of urban land use change in the Pearl River Delta, China: integrating remote sensing with socioeconomic data. Land Econ. https://doi.org/10.2307/3147108
Small N, Munday M, Durance I (2017) The challenge of valuing ecosystem services that have no material benefits. Glob Environ Chang. https://doi.org/10.1016/j.gloenvcha.2017.03.005
Smith JL (2008) A critical appreciation of the "bottom-up" approach to sustainable water management: embracing complexity rather than desirability. Local Environ 13(4):353–366. https://doi.org/10.1080/13549830701803323
Sosa-Rodriguez FS (2014) From federal to city mitigation and adaptation: climate change policy in Mexico City. Mitig Adapt Strat Glob Change. https://doi.org/10.1007/s11027-013-9455-1
State Council of China (2007) Notice on issuing the national environment and health action plan (in Chinese: 关于印发《国家环境与健康行动计划》的通知). Available at http://www.gov.cn/zwgk/2007-11/16/content_807439.htm
State Council of China (2013). National strategy for climate change adaptation (in Chinese: 国家适应气候变化战略). Available at https://www.gov.cn/gzdt/att/att/site1/20131209/001e3741a2cc140f6a8701.pdf
State Council of China (2014) National new-type urbanisation plan (2014-2020) 国家新型城镇化规划 (2014–2020年) (2014) (testimony of State Council of China). Available at: http://www.gov.cn/zhengce/2014-03/16/content_2640075.htm
Stephenson P (2013) Twenty years of multi-level governance: 'where does it come from? What is it? Where is it going?'. J. Eur. Public Policy 20(6):817–837. https://doi.org/10.1080/13501763.2013.781818 (Building better theoretical frameworks of the European Union's policy process))
Stone CN, Feagin JR (1990) Regime politics: governing Atlanta, 1946–1988. Contemp Sociol. https://doi.org/10.2307/2072466
Storbjörk S, Hedrén J (2011) Institutional capacity-building for targeting sea-level rise in the climate adaptation of Swedish coastal zone management. Lessons from Coastby. Ocean Coastal Manage. https://doi.org/10.1016/j.ocecoaman.2010.12.007
Strohbach MW, Arnold E, Haase D (2012) The carbon footprint of urban green space—A life cycle approach. Landsc Urban Plan. https://doi.org/10.1016/j.landurbplan.2011.10.013
Strohbach MW, Haase D (2012) Above-ground carbon storage by urban trees in Leipzig, Germany: analysis of patterns in a European city. Landsc Urban Plan. https://doi.org/10.1016/j.landurbplan.2011.10.001
Syrbe RU, Walz U (2012) Spatial indicators for the assessment of ecosystem services: providing, benefiting and connecting areas and landscape metrics. Ecol Ind. https://doi.org/10.1016/j.ecolind.2012.02.013

Tamosiunas A, Grazuleviciene R, Luksiene D, Dedele A, Reklaitiene R, Baceviciene M, Vencloviene J, Bernotiene G, Radisauskas R, Malinauskiene V, Milinaviciene E, Bobak M, Peasey A, Nieuwenhuijsen MJ (2014) Accessibility and use of urban green spaces, and cardiovascular health: findings from a Kaunas cohort study. Environ Health. https://doi.org/10.1186/1476-069X-13-20

Tao R, Su F, Liu M, Cao G (2010) Land leasing and local public finance in China's regional development: evidence from prefecture-level cities. Urban Stud. https://doi.org/10.1177/0042098009357961

Taylor N (1998) Urban planning theory since 1945. SAGE Publications. https://www.amazon.com/Urban-Planning-Theory-since-1945/dp/0761960937

Thomas MJ (1979) The procedural planning theory of A. Faludi. In: planning outlook. https://doi.org/10.1080/00320717908711582

Tompkins EL, Eakin H (2012) Managing private and public adaptation to climate change. Glob Environ Chang. https://doi.org/10.1016/j.gloenvcha.2011.09.010

Turner MG, Gardner RH (2015) Conclusions and future directions. In: Landscape ecology in theory and practice. Springer, New York, pp 383–396. https://doi.org/10.1007/978-1-4939-2794-4_10

Turner MG, Gardner RH, O'Neill RV (2003) Introduction to landscape ecology. In: Landscape ecology in theory and practice, 1st edn. Springer, pp 1–23. https://doi.org/10.1007/0-387-21694-4_1

Ulmer JM, Wolf KL, Backman DR, Tretheway RL, Blain CJ, O'Neil-Dunne JP, Frank LD (2016) Multiple health benefits of urban tree canopy: the mounting evidence for a green prescription. Health Place. https://doi.org/10.1016/j.healthplace.2016.08.011

UNDP (2010) Designing climate change adaptation initiatives. A UNDP toolkit for practitioners

UNFCCC (2015) UNFCCC adaptation calendar 2015—Building climate resilience in urban areas. https://unfccc.int/files/adaptation/application/pdf/calendar_final.pdf

UN-Habitat (2015) International guidelines on urban and territorial planning. https://unhabitat.org/sites/default/files/download-manager-files/IG-UTP_English.pdf

UN-Habitat (2016) Habitat III new urban agenda: quito declaration on sustainable cities and human settlements for all. In: Habitat III conference. http://habitat3.org/wp-content/uploads/NUA-English.pdf

United Nations. (2015a) Adoption of the Paris Agreement. In: Conference of the parties on its twenty-first session

United Nations (2015b) Summary of the Paris Agreement. In: United nations framework convention on climate change

van den Berg M, Coenen F (2012) Integrating climate change adaptation into Dutch local policies and the role of contextual factors. Local Environ. https://doi.org/10.1080/13549839.2012.678313

Vincent L (2018) MOOC: management of urban infrastructures. École Polytechnique Fédérale de Lausanne/Coursera. https://www.coursera.org/learn/managing-urban-infrastructures-1

Wang C-Y (2015) Between flexibility and reliability: changing planning culture in China. TU Delft. https://books.google.pt/books/about/Between_Flexibility_and_Reliability.html?id=CDQdCwAAQBAJ&redir_esc=y

Whittinghill LJ, Rowe DB, Schutzki R, Cregg BM (2014) Quantifying carbon sequestration of various green roof and ornamental landscape systems. Landsc Urban Plan. https://doi.org/10.1016/j.landurbplan.2013.11.015

World Health Organization (WHO) (2015) WHO global strategy on people-centred and integrated health services. In: WHO/HIS/SDS/2015.6

WHO Regional Office for Europe (2016) Urban green spaces and health—A review of evidence. https://www.euro.who.int/__data/assets/pdf_file/0005/321971/Urban-green-spaces-and-health-review-evidence.pdf?ua=1#:~:text=At.theFifthMinisterialConference,tokindergartensandschools%2Cand

WHO (2016a) Urban green spaces and health: a review of the evidence. World Health Organization

WHO (2016b) Health promotion: healthy China 2030 (from vision to action). Available at https://www.who.int/healthpromotion/conferences/9gchp/healthy-china/en/

WHO (2016c) Health promotion—Shanghai consensus on healthy cities 2016. Available at https://www.who.int/healthpromotion/conferences/9gchp/healthy-city-pledge/en/
World Health Organization (2018b) COP24 Special report: health and climate change
Wilby RL, Keenan R (2012) Adapting to flood risk under climate change. Prog Phys Geogr. https://doi.org/10.1177/0309133312438908
Wolch JR, Byrne J, Newell JP (2014) Urban green space, public health, and environmental justice: the challenge of making cities "just green enough." Landsc Urban Plan 125:234–244. https://doi.org/10.1016/j.landurbplan.2014.01.017
Wolfram M, Frantzeskaki N, Maschmeyer S (2016) Cities, systems and sustainability: status and perspectives of research on urban transformations. Curr Opin Environ Sustain. https://doi.org/10.1016/j.cosust.2017.01.014
World Bank (2017) World development report 2017: governance and the law. http://www.worldbank.org/en/publication/wdr2017
World Development Report (2010) Development and climate change. In: Carbon Dioxide Information Analysis Center. https://doi.org/10.3334/CDIAC/cli.002
WHO Europe (2017) Urban green spaces: A brief for action. P2
Wu F (2000) The global and local dimensions of place-making: remaking Shanghai as a world city. Urban Stud. https://doi.org/10.1080/00420980020080161
Xie L (2019) Contesting Chinese smart eco-cities: the promise of social justice and environmental well-being. In Zhang X (ed) Remaking sustainable urbanism: space, scale and governance in the new urban era, 1st edn. Springer, Singapore; Palgrave Macmillan, p. 240
Xu L, You H, Li D, Yu K (2016) Urban green spaces, their spatial pattern, and ecosystem service value: the case of Beijing. Habitat Int 56:84–95. https://doi.org/10.1016/j.habitatint.2016.04.005
Yin H, Kong F, Zong Y (2008) Accessibility and equity assessment on urban green space. Acta Ecol Sin 28(7):3375–3383
Zhang J, Sui YH, Geng XB (2011) Landscape design of urban green space adaptive to global climate change a review. Adv Mater Res. https://doi.org/10.4028/www.scientific.net/AMR.243-249.6842
Zhang Y, Murray AT, Turner BL (2017a) Optimizing green space locations to reduce daytime and nighttime urban heat island effects in Phoenix, Arizona. Landsc Urban Plan 165(March):162–171. https://doi.org/10.1016/j.landurbplan.2017.04.009
Zhang B, di Xie, xi Gao J, Yang Y (2014) The cooling effect of urban green spaces as a contribution to energy-saving and emission-reduction: a case study in Beijing, China. Build Environ. https://doi.org/10.1016/j.buildenv.2014.03.003
Zhang Y, Murray AT, Turner BL (2017b) Optimizing green space locations to reduce daytime and nighttime urban heat island effects in Phoenix, Arizona. Landsc Urban Plan. https://doi.org/10.1016/j.landurbplan.2017.04.009
Zhang T (2008) Developing the Chinese planning theory—Planning theory in the age of diversity. In: Conference paper "ACSP—AESOP—Joint congress."
Zhao Y (2005) Urban planning in the perspective of institutional economics. City Plan Rev 29(06&07):40–47, 17–27. http://en.cnki.com.cn/Article_en/CJFDTOTAL-CSGH200506008.htm
Zhu J (2008) Urban development under ambiguous property rights: a case of China's transition economy. Int J Urban Reg Res 26(1):41–57. https://doi.org/10.1111/1468-2427.00362
Zhu D, Zhou L (2015) Crisis and response: the transformation of chinese environmental governance institutional framework in contemporary China. Fudan J (Soc Sci) 180–188
Zürn M (2012) Global governance as multi-level governance. In: Levi-Faur D (ed) The Oxford hnadbook of governance (Illustrate). Oxford University Press. https://doi.org/10.1093/oxfordhb/9780199560530.013.0051

Chapter 3
Theoretical Framework

3.1 Semantics Matter

So far, key concepts such as *urban green spaces*, *governance*, *institutions*, *social costs,* and relevant literature have been reviewed without carefully examining the semantics. It is necessary to zoom into the semantics at the beginning of constructing a theoretical framework, as they matter for the empirical case study analysis, which is context-sensitive politically and culturally. Table 3.1 presents several examples of concepts, the meanings of which differ in English and Chinese. For instance, in Chinese, to *govern*: 治理 (pronounced as **z**hì lǐ) means "to treat and to organise," which can be used not only in social or political terms but also in many other fields, including medical and agricultural fields, a disease, or a type of pests. In comparison, in English, the term to govern applied more strictly in social or political terms and targets exclusively items, structures, and forms created by humans.

Another example is the term urban green spaces, the focal object of this research, which refers mainly to its functions as services. While the term in Chinese: 城市绿地 (chéng shì lǜ dì) means "urban green land," which emphasises its land property feature. A third example is the term *institutions*, which could be used to refer to an established/formalised practice. The corresponding term for *institutions* in Chinese is 制度 (zhì dù), which means "a crafted measurement, degree, or extent." One could argue that the reference to *institutions* in Chinese contains greater meticulousness. One might have similar observations on the term *citizens*, which corresponds to four different terms in Chinese: 公民 (gōng mín)—a person who has the legal right to belong to a particular country, 居民(jū mín)—residents, 市民(shì mín)—people who live in a city, and 公众 (gōng zhòng)—public audience. This brief showcase of semantics serves as a teaser of contextualisation in this book and a foundation for building the theoretical framework.

© Zhejiang University Press 2023
J. Liu, *Transforming Urban Green Space Governance in China Under Ecological Civilization: An Institutional Analysis*, Urban Health and Wellbeing,
https://doi.org/10.1007/978-981-99-6694-3_3

Table 3.1 Semantics matter: examples of key terms for this research and their meanings in English and in Chinese

Term in English	Meaning	Term in Chinese and Pronunciation	Meaning
Govern	To control and direct the public business of a country, city, group of people, etc.	治理**zhì lǐ**	To treat and to organise
Costs	1. The amount of money that you need to buy, make, or do something	成本**chéng běn**	The **foundation** needed to establish (something)
	2. The effort, loss, or damage that is involved to do or achieve something	代价**dài jià**	The **sacrifice** made to achieve (something)
Urban green spaces	A component of "green infrastructure" and an important part of public open spaces and common services provided by a city	城市绿地**chéng shì lü dì**	Urban green **land** (property)
Institutions	1. An established/ formalised practice	制度**zhì dù**	A crafted measurement, degree/extent
	2. An organisation founded for a certain purpose	体制**tǐ zhì**	A social, economic, or political system
Citizen	1. A person who has the legal right to belong to a particular country	公民**gōng mín**	=
	2. A person who lives in a particular place	居民**jū mín**	Residents
	/	市民**shì mín**	People who live in a **city**
	/	公众 **gōng zhòng**	Public audience
Civilisation	A state of human society that is very developed and organised	文明**wén míng**	The total wealth created by human beings, especially spiritual wealth, such as literature, art, education, science
	A society, its culture, and its way of life during a particular period of time or in a particular part of the world		

3.2 Overview of the Theoretical Framework

Having established the semantic foundation of key concepts used in this research, a theoretical framework outlining the essence of Urban green spaces as part of the urban social-ecological systems and centralising the common-pool resource theory is created to understand three empirical Urban green spaces governance cases in

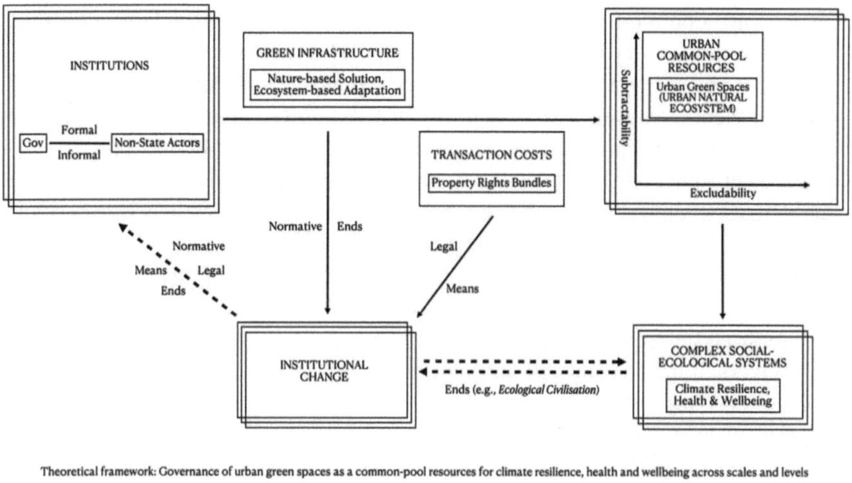

Fig. 3.1 Theoretical framework: Governance of urban green spaces as a common-pool resources for climate resilience, health, and well-being across scales and levels

the natural, social-economic, political, and cultural context of Guangzhou, China, as shown in Fig. 3.1. In a nutshell, Urban green spaces are an essential part of the urban social-ecological systems, the stock and flows of which are critical for the system's overall stability, including climate resilience and health and well-being. The stock and flows of urban green spaces are best governed as a common-pool resource considering its high subtractability and low excludability. The property rights arrangements among the actors of Urban green spaces governance influence the transaction costs, which affect actors' interactions, ultimately leading to formal and informal changes in Urban green spaces institutions.

The meanings of each box of the theoretical framework (see Fig. 3.1) are as follows:

- The upper left box shows that governments and non-state actors relate and interact with each other or within themselves through formal and informal institutions, as reasoned in Sects. 3.4.3.1 Governance and actors, and 3.4.3.2 Institutions.
- The upper middle-left box shows that green infrastructure (GI) includes Nature-based Solutions (Nature-based Solutions), which includes Ecosystem-based Adaptation (Ecosystem-based Adaptation), as reasoned in Sect. 3.3 Urban Social-Ecological Systems.
- The upper middle right box shows that property rights bundles constitute transaction costs. The way different property rights are bundled together influences transaction costs, as reasoned in Sects. 3.4.3.3 Property rights bundles and 3.4.3.4 Transaction costs.
- The upper right box shows that urban green spaces (Urban green spaces) are a type of urban common-pool resources (CPR), whose excludability is low, and the subtractability is high. As reasoned in Sect. 3.4.2 Why urban green

spaces are common-pool resources, urban green spaces provide a wide range of ecosystem services for all urban residents. The subtractability of Urban green spaces increases as urban population density grows, and land-driven economic development advances. The availability of Urban green spaces depends upon the total amount of urban land, which often has a geographical boundary.
- The lower right box shows that the status of complex ecological systems is dependent on its level of urban health and well-being and climate resilience, as reasoned in Sect. 3.3 Urban Social-Ecological Systems.
- The lower left box shows institutional change. The types of institutional change (*means* and *ends*) and the mechanisms (e.g., *normative, legal, formal, informal*) are reasoned in Sect. 3.4.3.5 Institutional change.
- The layers of the boxes mean different *levels* and *scales*, see reasonings about the *levels* and *scales* of institutions in Sect. 3.4.3.2 Institutions.

The causal relations between the boxes of this theoretical framework are as follows:

- The upper part shows that government and non-state actors interact and together govern Urban green spaces (which is a CPR) through applying Ecosystem-based Adaptation (a type of Nature-based Solution, which is based on utilising GI) and arranging property rights bundles (to manage transaction costs).
- The right part shows that the results of Urban green spaces governance or intervention influence the status of resilience to climate change, urban health, and well-being, thus influencing the stability/quality of the urban social-ecological systems.
- The lower part shows that the degrees of resilience to climate change, health, and well-being are end goals that contribute to institutional change.
- The middle right part shows that changing how property rights are bundled together constitutes legal changes in the institutions or the institutional change of the means.
- The middle-left part: Applying Nature-based Solutions contributes to achieving the end goal of climate adaptation and resilience enhancement. Changing how Nature-based Solutions are perceived is a normative or informal institutional change, as reasoned in Sect. 3.4.3.5 institutional change.
- The left part: institutional change might further influence the relationships between governments and non-state actors through legal or normative measures, which might change the means or the ends of the interactions between governments and non-state actors.

The next section outlines the essence of Urban green spaces as part of the urban social-ecological systems, in which trendy policy concepts such as GI, Nature-based Solutions, and Ecosystem-based Adaptation are also reexamined.

3.3 Urban Social-Ecological Systems

Urban green spaces, such as parks, woods and artificial meadows, ecological corridors, urban wetlands, and other natural ecosystems, are integral to the urban natural ecosystems. According to The Economics of Ecosystems and Biodiversity (TEEB) global initiative,[1] urban green spaces provide four essential ecosystem services: provisioning, regulating, supporting, and cultural services. Provisioning services include food, freshwater, and medicinal and raw material resources. Regulating services include local climate regulation, air purification, carbon sequestration and storage, moderation of extreme events, wastewater treatment, soil fixation, and pollination. Supporting services include habitats for species and maintenance of genetic diversity. Cultural services include recreation, mental and physical health, tourism, aesthetic appreciation and inspiration for culture, art and design, spiritual experience, and a sense of place. Generally, regulating and cultural services play a more significant role for cities than providing or supporting services (Larondelle et al. 2014).

The work which involves ecosystem services in research, policy, and practice spheres has been popularly referred to through various terms, such as green infrastructure (GI), nature-based solutions (Nature-based Solutions), ecosystem-based adaptation (Ecosystem-based Adaptation) in recent years, particularly in urban climate change and sustainability governance. However, they have mainly been applied with a utilitarian, human-nature dichotomic gesture (Welden et al. 2021). The European Environment Agency defines GI as "a strategically planned network of natural and semi-natural areas with other environmental features designed and managed to deliver a wide range of ecosystem services" in both rural and urban settings.[2] ICLEI—Local governments for sustainability defines Nature-based Solutions as "actions to protect, sustainably manage, and restore natural or modified ecosystems that address societal challenges effectively and adaptively, simultaneously providing human well-being and biodiversity benefits." (ICLEI 2017). In the Nature-based Solutions Initiative of the University of Oxford, Nature-based Solutions is defined as "actions that work with and enhance nature to help address societal challenges."[3] While the IPCC Assessment Report 5 *Climate Change 2014: Impacts, Adaptation, and Vulnerability* lists many types of Ecosystem-based Adaptation options, including soil conservation, ecological restoration, afforestation and reforestation, GI and ecological corridors, fisheries control and co-management, seed and gene banks (IPCC 2014). The conceptual scope of these three terms is: GI > Nature-based Solutions > Ecosystem-based Adaptation, as GI refers broadly to all plant species and the subecosystems that thrive around them, while Nature-based Solutions focus on instrumentalising specific subecosystems to solve problems. Ecosystem-based

[1] http://www.teebweb.org/resources/ecosystem-services/

[2] European Commission (2013).

[3] The Nature-based Solutions Initiative: *What are Nature-based Solutions?* Available at: https://www.naturEcosystem-based-Adaptationsedsolutionsinitiative.org/what-are-nature-based-solutions/

Adaptation is even more specific. These definitions and references share the common wish for yielding the most benefits from the ecosystems possible to tackle natural or man-made environmental challenges and serve people better, which shows an instrumental or technocratic approach to various extents. Besides, some scholars argue that the definition of Nature-based Solutions is rather vague, overarching with other similar concepts, and therefore impractical (Nesshöver et al. 2017), and that the ecosystem services that the Nature-based Solutions concept is based on are not always beneficial (e.g., human allergic reactions to pollens and insects) (Lyytimäki 2013).

These terms recently gained popularity, particularly in urban policy and practice. The C40 City Climate Leadership Group endorses the GI approach for climate change adaptation. The emerging concept *of Sponge City* and the recent application of nature-based floodwater treatment schemes in Chinese cities is one type of Nature-based Solution. It means the city can be like a sponge and has good "flexibility" in adapting to environmental changes and responding to natural disasters. Absorb, store, seep, and purify water occasionally, and "release" the stored water when needed and use it. *Sponge City* resonates with two other concepts—GI and Low Impact Development (LID) technology (Jia et al. 2017). LID refers to engineering practices that "use or mimic natural processes that result in infiltration, evapotranspiration or use of stormwater to protect water quality and associated aquatic habitat."[4] The Global Assessment Report produced by the Intergovernmental Science-Policy Platform on Biodiversity and Ecosystem Services (IPBES) has also endorsed urban green spaces as a possible pathway to transform urban biodiversity and as an urban nature solution to climate change impacts.

While these terms might be limited in various ways, they share a common intention to operationalise complex ecological systems to address these interconnected environmental problems for greater social well-being in less artificial manners. Essentially, urban areas are hot spots of complex and dynamic interactions across ecological, social, cultural, and economic systems, which scholars have referred to as "urban social-ecological systems (SES)" (Elmqvist et al. 2019; Frank 2017; Li et al. 2014; Liu and Gatzweiler 2020; McPhearson et al. 2016a, b). This definition is more holistic than the previously mentioned ones—GI, Nature-based Solutions, and Ecosystem-based Adaptation, and conveys the image of nature-human coexistence as integrated, complex living organisms (Kuzevanov and Gubiy 2014). Urban green spaces are part of the natural ecosystem, which, together with artificial ecosystems, compose an urban social-ecological system. As Fig. 3.2 shows, if managed by collaborative agents (people and organisations) through *integrative institutions*,[5] a diverse range of services produced by both the natural and artificial ecosystems will together contribute to enhance the climate resilience and human well-being of

[4] Environmental Protection Agency (EPA) (n.d.a.) "Urban Runoff: Low Impact Development" Available at: https://www.epa.gov/nps/urban-runoff-low-impact-development.

[5] According to Hagedorn's (2008, 2015) "Institutions of Sustainability" theory, institutions can also be distinguished as *integrative institutions* and *segregative institutions*. *Integrative institutions* are the ones that can reduce unintended externalities that burden the health and well-being of vulnerable groups in society or the environment. For more, please see Sect. 3.4.3.5 *Institutional change*.

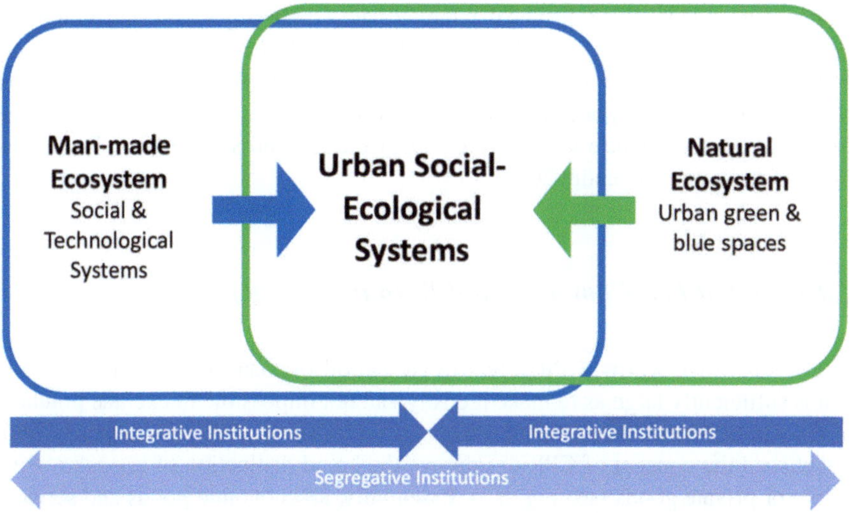

Fig. 3.2 Urban social-ecological systems and effects of integrative and segregative institutions. *Source* Adapted from Liu and Gatzweiler (2020)

the entire urban social-ecological system while *segregative institutions* would do the opposite (Armson et al. 2012; Hartig et al. 2014; Panno et al. 2017; Tyler and Moench 2012; Zölch et al. 2017). Understanding the essence of Urban green spaces as part of the complex urban social-ecological systems is critical to see where Urban green spaces stay a resource in the entire system, how its stock flows within the system, and how to manage it.

3.4 The Common-Pool Resource (CPR) Theory

This section zooms in to establish the theoretical foundations for understanding how to manage urban green spaces (Urban green spaces) as a resource as a common-pool resource (CPR). It is composed of four parts:

1. What is a common-pool resource (CPR)?
2. Why Urban green spaces are CPRs?
3. Critical elements to be considered in CPR governance, and
4. an analytical framework to analyse how these elements determine CPR governance in empirical settings.

In a nutshell, the CPR theory contradicts Hardin's "tragedy of the commons" model and convenes that: compared to state-dominated public or market-oriented private approaches, goods that are non-excludable to community members but are rivalrous in consumption can be governed more sustainably in small scales and through collectively designed rules including adequately arranged property rights, monitoring, and sanctions, despite that humans are supposedly selfish and boundedly rational (Ostrom 1990, 2005; Schlager 2004). The CPR theory is developed mainly by Nobel Economics Laureate of 2009, Professor Elinor Ostrom, together with her collaborators, backed by decades of empirical research on resource governance across contexts (e.g., see Chhatre and Agrawal 2009; Coleman and Steed 2009; Gibson et al. 2005; Ostrom and Nagendra 2006).

3.4.1 What Is a Common-Pool Resource (CPR)?

In Ostrom's own words, CPR refers to "a natural or man-made resource system that is sufficiently large as to make it costly, but not impossible to exclude potential beneficiaries from obtaining benefits from its use" (Ostrom 1990: 30). Ostrom (1990) established the concept of CPR to improve the broad distinction of goods as being public or private goods (see Fig. 3.3). Essentially, all economic goods and services can be distinguished by excludability and subtractability. Excludability refers to the quality of excluding users from consuming certain goods. In contrast, subtractability refers to the quality of rivalry for one user vs for other users, or in other words, the impact of the consumption by one user on the quantity available for other users. Public goods, also referred to as "social goods" or "collective consumption goods," have low excludability and subtractability (Bowen 1943; Musgrave 1939; Samuelson 1954, 1955). This means, the cost of extending the service of such goods to an additional person is zero, and it is impossible or expensive to exclude individuals from enjoying, or can only exclude them at a very high exclusion cost (Nordhaus 2009). Air, water, and soil are typical examples of public goods. Peace and security of a community are also public goods. Private goods, on the contrary, have high excludability and high subtractability. This means the governance options are rather limited to individuals or groups that own them, with little possibility for governance disputes.

Essentially, the goal of governing public goods is to achieve a balance between the supply and demand ends of goods and services. This balance builds on four assumptions: (1) perfect information flow, (2) zero opportunity costs, (3) zero stock externalities, and (4) rational decision-making (Nordhaus 2009; Samuelson and Nordhaus 2010). However, neither state nor market governance approaches to public goods have always been successful in addressing social and environmental externalities due to the following factors:

1. *Information mismatch.* The *Laissez-faire* (French: "allow to do") market economy theory assumes buyers and sellers are completely informed about a transaction they make, but in fact, information flow in any real-world transactions, in general, is imperfect.
2. *Opportunity costs* refer to "the value of the most valuable good or service forgone" (Samuelson and Nordhaus 2010: 139). For example, deforestation of a forest area rich in biodiversity for timber bears the opportunity cost of missing the carbon

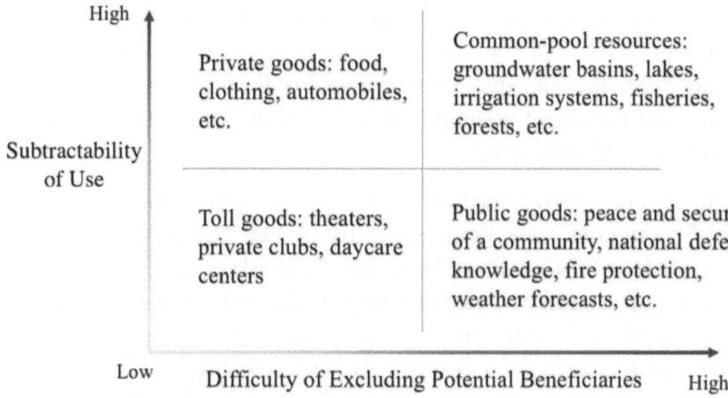

Fig. 3.3 Four basic types of goods. *Source* Adapted from Ostrom (2005: 24)

sequestration and biodiversity maintenance services the forest offers. A growth-oriented and high-emission urban development model bears the opportunity costs of public health and well-being, quality of the urban environment, and urban biodiversity.

3. *Stock externality* refers to the impact of negative externalities accumulating over time (Samuelson and Nordhaus 2010). For example, coastal land resources are a good example. Coastal erosion due to sea-level rise and intensified storms results from global warming. They negatively affect the quality and quantity of coastal ecosystems and populations, accumulating slowly. The consequences of coastal erosion—loss of coastal biodiversity and human settlements—are often long-lasting and maybe even irreversible. The stock externalities of public goods also project uncertainties into time, spatial, and political dimensions (Nordhaus 2009). On the time dimension, for example, it is nearly impossible to estimate the impact of global warming on coastal land resources by 2100 accurately. On the spatial and political dimensions, for example, it is challenging to estimate the impact of China's CO_2 emission on the potential flash flood damages in the Himalayas. National governments have the legal and political powers to arrange institutions to internalise negative externalities, but only within their spatial/territorial boundaries.
4. *Bounded rationality* (Ostrom 1990; Samuelson and Nordhaus 2010). Arrow (1974: 16) argues that economists often "…thinks of himself as the guardian of rationality, the ascriber of rationality to others, and the prescriber of rationality to the social world." In reality, economists and government officials make judgements and decisions the same way as any lay people in imperfect conditions that prevent them from thinking things through, such as constraints of earlier choices, undiscovered, ignored, or misunderstood facts (Rothman 2019). Simon (1956, 1990) refers to these imperfect human cognitive conditions as *bounded rationality.* Both market and state approaches tend to function with rational assumptions about information flow and propensity for efficiency, leading to

the spillover of externalities (Chambwera et al. 2015; Kostka 2015; Samuelson and Nordhaus 2010). The failures to govern many natural resources sustainably largely are due to the inadequate consideration of the high subtractability of many resources, and the above assumptions taken by the conventional state-oriented or market-oriented approaches.

Although Ostrom (1990) came to create the term common-pool resources (CPR), the notion of "commons" was first defined by Garrett Hardin through his notable essay *The Tragedy of the Commons.* Nonetheless, Hardin's attempt has been widely criticised for being simplistic and proposing market-oriented solutions that could not mitigate social and environmental externalities. Compared to public goods such as peace and security of a community, or knowledge, CPRs have the same non-excludability feature as public goods. However, unlike public goods, the subtractability of CPRs is high (again, see Fig. 3.3). Ostrom distinguishes many non-excludable but rivalry goods, such as forests, water systems, fisheries, and the global atmosphere, as CPRs. Once used by one user, these natural resources inevitably become less available for others, yet, they are of immense importance for human survival on Earth.

Furthermore, both public goods and CPRs face the problems of free-riding or opportunism. The costs of excluding other potential users for public goods and CPRs are high (Ostrom 1990). Despite facing similar free-riding issues, CPRs are more precarious than public goods, as the subtractability of CPRs is lower than that of public goods. Studies on the commons emerged in the mid-1980s as the International Association for the Study of the Commons was established to dedicate to this rising subject. The field gained accelerated interest, particularly as Ostrom won the Nobel Prize in Economics in 2009. Since then, studies using the *Commons* lens have propagated from the initial focus on the environment or natural resources, such as forestry and irrigation systems, to various items such as conservation sites, climate change, and many other non-traditional CPRs (Wang 2017). The *commons* perspective in this book is particularly relevant in the context of global climate change impacts and sustainability challenges, which are complex and uncertain, with diffused benefits which discount in the long-term and transcend sovereign boundaries.

3.4.2 Why Urban Green Spaces Are Common-Pool Resources?

Urban green spaces can be considered a type of CPRs crucial for urban health and climate resilience. As Wendell Berry (1987) portrays: "*... is a commonwealth: a place, a resource, an economy. It answers the needs, practical as well as social and spiritual, of its members – among them the need to need one another.*" Viewing Urban green space resources from the *Commons* perspective responds to the big picture in which the needs of the current generation and those of future generations are considered. Urban green spaces, such as urban parks, wetlands, ecological corridors, and street trees, have low excludability and high subtractability characteristics. Urban green space users are residents who use them for physical exercises, better air, water and soil quality, contact with nature, and mitigation of the impacts of extreme weather such as heatwaves and urban inundation. Urban green space users could also be urban land users, such as the local government, enterprises, and urban residents. Urban green spaces are a non-excludable resource providing ecosystem services for all urban residents. The subtractability of Urban green spaces increases as urban population density grows and land-driven economic development advances. The availability of Urban green spaces depends upon the total amount of urban land, which often has a geographical boundary.

Similarly, urban land also shares the characteristics of CPR. Economists deem land resources as the single most valuable natural resource for their scarcity and significance for the production of almost all businesses (Samuelson and Nordhaus 2010). Urban land has a pronounced scarcity, which makes it essential to produce goods and services of adequate quantity and quality within available urban land spaces, including vital ecosystem services.

Urban infrastructures are increasingly considered strategic arenas for climate change action (Broto 2017). Within this context, Urban green spaces are increasingly conceptualised as a CPR globally and the term "urban green commons" emerged from the perspective of property rights to address the governance issues of Urban green spaces within various land ownership conditions (private, public, club, or hybrid) (Colding et al. 2013) In China, the knowledge gap of CPRs is even more evident. **Despite being a country with highly diverse and abundant CPRs, including watershed, forests, environment, regional ecosystems, and governance practices in the world, scientific research from the perspective of commons in China or about Chinese CPRs are lacking, and much less has been communicated within the international scientific community** (Wang 2017)**.**

3.4.3 Critical Elements in Common-Pool Resource Governance

3.4.3.1 Governance and Actors

The concept of *governance* has come to be defined more comprehensively in literature each time. The Commission on Global Governance (1995) broadly defines *governance* as the sum of how individuals and institutions, public and private, manage issues. While the Natural Resource Governance Institute (2017) works with the definition of *governance*, emphasising the role of state actors: "the form of political regime or how authority is exercised in the management of a country's social or economic resources for the public good." The World Bank (2017) defines *governance* more comprehensively as "the process through which state and non-state actors interact to design and implement policies within a given set of formal and informal rules that shape and are shaped by power—the ability of groups and individuals to make others act in the interest of those groups and individuals and to bring about specific outcomes." Similarly, the UN-Habitat sourcebook on *Integrating Health in Urban and Territorial Planning* defines governance comprehensively as "the political and administrative management of places that involves partnerships (formal or informal) between governments at different levels but also the private sector and civil society organisations" (Grant 2020).

More broadly, *governance* is the process and outcome of decision-making because of interactions between state and non-state actors according to the rules. State actors refer to governments of different administrative levels. American economist Arrow (1974: 25) defines *government* as "one of a large number of collective institutions which distinguishes itself from others primarily by its monopoly on coercive power, although even that monopoly is not absolute." State actors seek rent by providing a series of services, such as protecting property rights, providing public services, and military defence. In this perspective, state actors, to their best capabilities, make policies based on pragmatic calculations of costs and benefits in economic, social, environmental, and political spheres. Non-state actors, conversely, encompass individuals, associations, ethnic groups, organisations, and institutions across regional and local levels that are not governments. Non-state actors influence state decisions directly through interactions with states or indirectly through the dynamics between each other, including information sharing, capacity building, or even competing (Andonova et al. 2009). Non-state actors play a significant role in urban planning and environmental governance across levels (Biermann et al. 2010). Non-state actors—urban planners, developers, and citizens—are also considered the pragmatic "economic man" who seek to maximise their benefit–cost ratios (environmental, social, or economic).

3.4.3.2 Institutions

Institutions are crucial for urban sustainability governance as they structure interactions within governments and between government and non-state actors by reinforcing norms and commitments, facilitating coordination, and promoting cooperation and implementation (Bridges 2016; Latour and Douglas 1988; World Bank 2017). One of the most widely cited definitions of *institutions* is by American economist Douglass North, Nobel laureate in Economics (1984): "(*institutions* are) contractual arrangements between principals and principals and agents, made to maximise their wealth by realising the gains from trade as a result of specialisation (including specialisation in violence or coercion)." In North's definition, an agent explicitly or implicitly gives up control over their decision-making power and submits authority to a principal. For example, a worker to an employer or manager and a bureaucrat to a government official are relationships of agents to principals. North defined *institutions* as "the humanly devised constraints that structure political, economic and social interaction." North (1991) distinguishes institutions as containing both informal institutions, such as strategies, ethics, norms, customs, sanctions, taboos, traditions, and codes of conduct, and formal institutions, including constitutions, laws, property rights, legal statements, and policies." Hodgson (2006) puts *institutions* more simply as "systems of established and embedded social rules that structure social interactions." Ostrom defines *institutions* as "the prescriptions that humans use to organise all forms of repetitive and structured interactions including those within families, neighbourhoods, markets, firms, sports league, churches, private associations, and governments at all scales." (2005: 3). According to Ostrom, institutions exist whenever interdependent individuals act in an organised way; their high complexity results from the diverse situations in contemporary life. The variations in levels, scales, and types of human-organised social life make institutional arrangements, or the game's rules, highly complex.

Reasonably designed, functional institutions can structure human behaviour and actor interactions and generate efficiency in a long-lasting way (Vatn 2005). Acemoglu and Robinson, the authors of *Why Nations Fail* (2012), explained the origin of the rise and fall of nations through the study of various countries throughout different eras. They argue that institutions of a country can directly led to its prosperity or turmoil. *Functional or inclusive institutions* can promote the research and development of new technologies and are the foundation of sustainable economic development. *Extractive institutions*, however, even if they promote economic growth, can only lead to a temporary improvement in efficiency, which is difficult to sustain. In urban planning, it is not uncommon to see technically sound planning ideas fail to be implemented and sustained in practice, as planners often neglect or it is out of their professional reach to make institutional adaptations accordingly (Zhao 2005).

Besides, the dimensions of *levels and scales* can fundamentally impact the effectiveness of institutional arrangements. Due to the complexity of institutions and actors in broader field settings, analysis of urban environmental governance necessitates understanding interactions among actors and the impacts of institutional arrangements on them across these two dimensions (Ostrom 2010). The *levels* of institutions are generally related to the structures, administrative or legislative, through which governments organise non-state actors to participate in social and economic activities and interact with the environment (Cash et al. 2006). Administrative levels include nation, province, and local levels such as city, district, town, township, and village. Legislative levels include constitutions, laws, regulations, and context- or case-specific operating rules. The *scales* of institutions are commonly related to the dimensions of time, geography, population, and quantities of goods and services. Gibson et al. (2000) define *scale* as "the spatial, temporal, quantitative, or analytical dimensions used to measure and study any phenomenon" and level as the unit of analysis that is "located at the same position on a scale." For example, climate change impacts and biodiversity loss can appear locally but extend globally through cause-and-effect chains and transcend existing administrative-territorial jurisdictions or levels (Newig and Fritsch 2009).

Besides, institutions of resource use or environmental governance at local levels, in general, are embedded in and influenced by institutions at regional, national, and global levels. Increases in administrative levels and geographic scales lead to increased complexity and uncertainty, both in ecological conditions and governance operations (Olsen et al. 2011). In climate adaptation, multi-level institutional coordination between different political and administrative levels in society is increasingly challenging in developing and developed countries (Glaas and Juhola 2013; Rodima-Taylor et al. 2012). Establishing a multi-level institutional coordination mechanism between different political and administrative levels is crucial for promoting urban adaptation planning and implementation outcomes (Engle and Lemos 2010; Glaas and Juhola 2013; Revi et al. 2014). In addition, building institutional capacity at the local level, e.g., in local governments, is critical too (Mimura et al. 2015).

Ostrom (2015) summed up eight common characteristics (in Ostrom's original words, "design principles") in long-enduring CPR governance regimes. Most successful CPR governance cases have:

1. Clearly defined boundaries, including resource boundaries and user boundaries,
2. Proportional equivalence between benefits and costs for all actors (individuals or groups),
3. Collective-choice or public participation arrangements,
4. Effective mechanisms of monitoring,
5. Graduated sanctions of actors who violate the rules collectively agreed upon,
6. Effective conflict resolution mechanism via affordable means,
7. Minimal recognition (by state actors, e.g., upper-level governments) of the rights to organise, and lastly,
8. Nested enterprises in which the resources in governance are considered part of a larger ecosystem or the governance activities are integrated into a multi-level governance scheme.

Over the years, economists have concluded that the market-oriented approach is incapable of governing common-pool resources effectively, for it (1) generates negative externalities in the process of goods and services distribution, (2) quantifies the costs and benefits of the diverse ecosystem services narrowly, which undermines ecological and community values, particularly of those socially and economically underprivileged changes constantly in economic value, and (3) constantly changes in economic value (Gatzweiler 2003; Samuelson and Nordhaus 2010; Chambwera et al. 2015). Besides, the model of the liberal economy often proclaims that technological means (e.g., direct air carbon capture and storage, abbreviated DACCS) and economic progress can overcome biophysical limitations of the Earth or any reckless, exploitative, or even violent human behaviours.

Thus, the role of public authorities gradually gained significance in natural resource governance for their potential to regulate the market-driven distribution of goods and services to reduce uncertainties, inequities, and behavioural biases (Chambwera et al. 2015). Public authorities are considered pragmatic and self-interested organisations, which are no exception but a member of the numerous actors in the market whose motives and behaviours obey the rules of the market game and gain the maximum surplus (Zhao 2005). Arrow considers authority as "the centralization of decision-making" and can serve "to economizes on the transmission and handling of information" (1974: 69). For example, a direct cost–benefit focused approach would assess the risks of flooding related to intensified tropical cyclones and storm surge in dense urban areas through the direct costs of damaged infrastructure, interruption of productivity, and hospitalisation. A progressed, state-supervised version of disaster impact assessment could go one step further to consider the social costs of the underprivileged communities and the indirect costs of recovery induced in the foreseeable future. However, even a more progressed, state-supervised assessment could not provide 100% of information certainty or availability over time due to the changing dynamics in the complex social-ecological systems. As examined more recently with community-based conservation, simple policy prescriptions to entrust resources to a state actor, a free market, or even to decentralised user groups may also fail if institutional complexity of levels, scales, objectives, actors, and the social contexts are not appropriately taken into account (Berkes 2007; Meinzen-Dick 2007).

3.4.3.3 Property Rights Bundles

Two major problems are associated with CPR governance: overexploitation and free-riding. Overexploitation occurs when the difficulty of excluding other beneficiaries is high. Free-riding happens when those who benefit from the CPRs do not pay for them, which results in under-provision of those resources. For example, a well-constructed by villagers to fetch water has costs of construction and maintenance. When some villagers want to enjoy the sweet water but do not want to share the maintenance cost with other villagers, the well could go into a mal-maintained state and eventually dry up.

These problems can be resolved by defining and clarifying *property rights* (Coase 1992; Ostrom et al. 1961). Alchian (1989) defines *property rights* as socially enforced rights to select uses of an economic good. More specifically, North (1990) considers *property rights* as rights that individuals or groups appropriate over their own labour and the goods and services they possess. Zhao (2005) argues that the essence of CPRs is not the exchange of goods and services itself but the exchange of a set of *property rights*. The value of the goods or services traded also depends on the range or the clarity of the property rights transferred in a transaction (Coggan et al. 2010; McCann et al. 2005). Demsetz (1974) and Coase (1998) consider that the prerequisite for all human interaction activities is institutional arrangements, a power to exercise certain behaviours between individuals or groups. Building on the same line of thought, Zhao (2005) argues that the primary tasks of institutional intervention are to define the property rights of individuals or groups to resources, stipulate what they can do, and then maximise the total social product through the transaction of their property rights.

Resources are often classified into different types, partially depending on their ownership and partially on other property rights. CPRs can be owned and managed as government, private, community properties or owned by no one (Bromley 1986). Through extensive meta-analysis of existing field cases, Ostrom (2005) and her colleagues have identified five types of property rights that individuals or groups using a CPR:

(1) Access—the right to enter a specified property,
(2) Withdrawal—the right to harvest specific products from a resource,
(3) Management—the right to regulate internal use patterns and to transform a resource,
(4) Exclusion—the right to decide who will have access, withdrawal, or management rights, and
(5) Alienation—the right to lease or sell any of the other four rights.

The concept of CPRs and these identified property rights is now widely accepted by scholars studying resource management issues in diverse systems worldwide (Degnbol and McCay 2007; Paavola and Adger 2005; Trawick 2001). Identifying and acting upon the right property rights are critical to success in resource management. As Zhao (2005) exemplified, many historical districts and traditional buildings with the best construction quality in Beijing deteriorated drastically within the last few decades, mainly due to the misplacement of property rights. Residents, entitled to access and maintain these buildings and in the best position to do so, were excluded from such rights. Property rights arrangements are critical for effective coordination and fair, efficient distribution of resources among the population for sustainability (Heltberg 2002).

Ideally, the property rights bundles in urban green spaces governance should be considered in two folds. The first is associated with the urban land resources, and the second is associated with the ecological resources—the ecological networks, the ecosystems, and the goods and services these ecosystems provide. In the case of private urban land properties, the rights to the land are mostly equivalent to those to the ecological resources that grow directly on it. For public land, ownership rights could belong to a public entity, such as a state organisation, but the ecological resources and the ecosystem goods and services on public land could be private or communal, and the rights to manage could be attributed further to other entities. Investigation into the property rights of Urban green spaces necessitates recognising the complex property rights bundles to (1) urban land, which could be public, private, or club lands (e.g., residential compounds), and (2) the ecological resources within the same geographical boundary. In the CPR theory, communities who are the most direct bearers of climate change consequences and access Urban green spaces for physical and mental health and well-being purposes, with the respective property rights attributed among actors in proper levels and scales, can organise an efficient, sustainable Urban green spaces governance regime. The institutional arrangements in urban planning are, in essence, distributions of property rights among actors. If property rights are well-defined, particularly when there are few affected parties, negative externalities could be avoided more easily (Samuelson and Nordhaus 2010).

3.4.3.4 Transaction Costs

In practical terms, economic activities involve costs in many processes to soother the political or economic relationships and interactions between individual or group actors, e.g., searching for and analysing information, reaching and implementing an agreement (Vatiero 2020). In many cases, the exact costs of a transaction are not easily quantifiable, but the concept of *transaction cost* informs the imperfection of transactions. Coase initially introduced the concept of *transaction costs* to explain the emergence of firms, or the internalisation of production of a particular good or service, as driven by the need to avoid or reduce costs included in the price of the good or service, such as costs of search and information acquisition, bargaining or negotiation, as well as enforcement, access to new markets, and many others (Coase 1937, 1960; Williamson 1979). Another institutional economist Douglass North defines *transaction costs* as "the costs of specifying and enforcing the contracts that underlie exchange and therefore comprise all the costs of political and economic organisation that permit economies to capture the gains from trade" (North 1984). North estimates the percentage of transaction costs in the gross national product (GNP) in advanced Western countries to be as much as 50%. The concept of *transaction costs* redirects the core principle of economic activities from obtaining the largest economic gains to reducing transaction costs. It helps organisations make decisions based on weighing the value of what is obtained and the value of what is sacrificed to obtain it and choose the economically efficient option (Chambwera et al. 2015).

Broadly speaking, when multiple actors or actor groups are involved in a decision-making setting, the *transaction costs* can also be viewed as the costs to achieve agreement or consensus; see Fig. 3.4. Situations A and B represent decision-making situations in which the costs to reach a consensus is assumed manageable. Situations C and D represent situations where the costs to achieve consensus are assumed beyond manageable. Situations A and C are when the differences of interests or information among actors or organisational members influence the costs more than the number of interests or information. In Situations A and C, there is more drive for authority and less room for consensus-seeking. In Situations B and D, the quantity of interests or information among actors or organisational members influences the costs more than the differences of interests or information. There is less drive for authority and more room for consensus-seeking in these circumstances. To promote actors in a decision-making setting to interact towards consensus-seeking, staying within the ideal quantity-versus-difference ratio of interests or information among actors. Situation B represents such a ratio. For Situation A, communication is needed to reduce the differences of interests or information for cost–benefit sustainable consensus-seeking. For Situation C, communication is necessary to minimise the differences in interests or information. Besides, downsizing the number of actors or organisational members is also required for cost–benefit consensus-seeking. For Situation D, downsizing is proposed to reduce the number of interests or information to make the consensus costs manageable.

State actors are pragmatic actors who also pursue the maximum benefits within environmental, social, and political spheres. At times when transaction costs are implicit or hard to detect, it is often due to the existence of hierarchies which are effective in allocating efficiency (Coase 1937). Due to the universality of uncertainty and bounded rationality, contracts, deals, and decisions are necessarily incomplete (Williamson 1977). In these occasions, hierarchy constructs a superior-subordinate

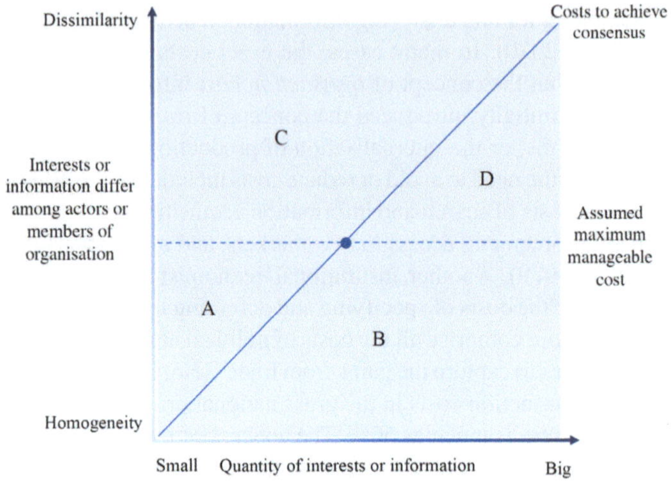

Fig. 3.4 Costs to achieve consensus dependent on the quantity and similarity of interests or information (and thereby power) between actors

relationship through which *transaction costs* are distributed or disseminated along with the downsizing of scales and levels. At least two actors, individuals, or groups are involved in transactions. This is also a prerequisite for institutions to emerge. Therefore, some scholars also refer to *transaction costs* as *institutional costs* (e.g., Cheung 1989). Ostrom (2010) and Schlager and Heikkila (2011) suggest that the key to reducing transaction costs among large groups of individuals when coping with complex social dilemmas is to communicate and make context-appropriate rules collectively.

In climate change mitigation and adaptation, a major transaction cost is information acquisition, e.g., the technical challenge to obtain updated climate and weather data or access to the existing one, or to quantify adequately the benefits of Urban green spaces for the ecosystem services they provide (Cimato and Mullan 2010). As a public good, information suffers the under-provision problem by private actors, hence requiring public authorities to support its procurement and dissemination, e.g., through research funding (Fankhauser et al. 1999; Trenberth 2008). *Adjustment cost* represents another barrier, especially in the presence of uncertainty and learning. Fankhauser et al. (1999) define *adjustment costs* as a barrier to early capital replacement to adapt to a different climate. Kelly et al. (2005) determine *adjustment costs* incurred while learning about new climate conditions. These definitions suggest that *adjustment costs* can represent a significant share of climate adaptation costs. *Adjustment cost* also includes defining the property rights to climate mitigation benefits and costs (Libecap 2014). In urban planning, analysis mainly focuses on two types of transaction costs; one is *technological cost*, e.g., when it involves engineering techniques or technologies, and the other one is *institutional cost due to the mismatch or imperfect flow of information*, as well as due to the credibility of agreements, or rules, such as property rights arrangements (Ho 2017; Zhao 2005). However, despite the theoretical value of *transaction cost*, how to measure it accurately for effective policy-making often remains a methodological challenge.

3.4.3.5 Institutional Change

Types and mechanisms of institutional change

Institutions are dynamic and evolutionary by nature, and they change over time. North (1990: 3) notes that, as institutions structure incentives in human interactions in political, social, and economic spheres, "institutional change shapes the way societies evolve through time and hence is the key to understanding historical change." *Institutional change* refers to the innovation and breakthrough of an institutional framework. *Institutional change* "typically consists of marginal adjustments to the complex of rules, norms, and enforcement that constitute the institutional framework" (North 1990: 83). Institutional change results from actors' interactions within given institutional arrangements over time. Different institutional arrangements expand or limit the choices and behaviours of actors.

Institutional change can be distinguished as *process-oriented* or *goal-oriented*, or in other words, institutional change of the *means* or the *ends*. The common-pool resource theory tends to focus on *process-oriented* institutional changes. Blomquist and Ostrom (2008) define *institutional change* as entailing institutional design, assessment, and modification and propose that it necessarily takes place within the constraints and opportunities afforded by existing institutional arrangements. North (1990) argues that the process of institutional change could be either *discontinuous* or *incremental*. While a *discontinuous change* of formal institutions might occur overnight because of political decisions, *incremental change* often refers to informal institutions such as tradition and cultural norms. In addition, depending on the actor(s) who induce the process of change, institutional change could be distinguished as *bottom-up* or *top-down* (North 1991). *Bottom-up* institutional change occurs when the change is induced by demand, e.g., changes that are initiated, organised, and realised by individual or group actors as they are attracted by the opportunities to profit from a potential new institution. While *top-down* changes occur when the change is induced by supply, e.g., when authorities introduce changes in the form of regulations and laws (Wegerich 2001).

One example of institutional change of the *means* that is *process-oriented* is the governance shift towards *polycentric governance* from state- or market-oriented approaches. The quest of polycentric governance focuses on the means—how to approach a subject, a resource, or a service, to achieve higher efficiency in all the desired dimensions. Institutional change of the means does not change the broader institutional context, or deeper lying institutions which define the ends, e.g., fundamental values or principles. The concept of *polycentric governance* builds on the premise that social-ecological systems, e.g., how people organise resource use of various scales across local, regional, and national levels, are becoming increasingly interconnected and complex (Berkes 2006; Blomquist 2009). Ostrom et al. (1961) and Ostrom (2005) define the concept of *polycentricity* as connotating multiple centres of decision-making that are formally independent and can function together within interdependent systems depending on contexts or cases. *Polycentric governance* emphasises the collective gesture of community organisation and the dispersed in many small groups or communities surrounding access of goods and services (Nagendra and Ostrom 2012). Another example of process-oriented institutional change of the *means* is the people-centred health approach endorsed by the World Health Organization (WHO), which "consciously adopts the perspectives of individuals, families and communities, and sees them as participants as well as beneficiaries of trusted health systems that respond to their needs and preferences in humane and holistic ways." (WHO et al. 2015: 10–11). According to the WHO, people-centred health services need to reflect a set of core principles, such as comprehensiveness in responding to people's evolving health needs and aspirations, equity, sustainability, prevention, empowerment, and systems-thinking. The WHO emphasises "the right care at the right time in the right place" (WHO et al. 2015: 14). The definition by the WHO for its member states in health service provision, hence, calls for institutional change primarily on the *means* of public service. The *means* is the inclusion of citizen's rights both to health benefits and to participation.

The other broad type of institutional change is *goal-oriented*, or institutional change of the *ends*. This type of institutional change focuses on the end goals and objectives more so than on the process or the means to achieve certain change. For example, according to Hagedorn's (2008, 2015) "Institutions of Sustainability" theory, institutions can also be distinguished as *integrative institutions* and *segregative institutions*. *Integrative institutions* are the ones that can reduce unintended externalities that burden the health and well-being of vulnerable groups in society or the environment, such as pollution, extinction, and degradation (McPhearson et al. 2016a, b). In this circumstance, the degree of *institutional fit* for the end goal (such as urban health, climate adaptation) is high (Epstein et al. 2015; Young 2002). *Integrative institutions* can evolve towards *segregative institutions* or *institutional mismatch*, when external environments change, resulting in changes in the end goals and objectives, thus stimulating institutional change. Neale (1993) considers that institutional change occurs when actors learn, adapt, and respond accordingly by putting forward demands for new institutions to achieve the expected increased benefits. Only when the expected benefit is greater than the expected cost, actors will make adaptations or changes to existing institutions (North and North 2012; Zhao 2005).

In some cases, the *people-centred* approach may also focus on the *ends* of development. According to Robertson (1994), a member of the People-Centred Development Forum, people-centred development is one that facilitates the transition to a new civilisation a sharp contrast to more familiar forms of development that have taken their mission to be the Euro-American civilisation and modernisation. In this Euro-American development model, wealth is augmented by ever-increasing exploitation of the Earth, governed by impersonal systems, and possesses greater legitimate power than people. According to Korten (1987), *people-centred* development strategies emphasise the need to strengthen institutional capacity to support greater local control, accountability, and self-reliance regarding access to physical and financial resources. The *people-centred development* defined by the People-Centred Development Forum is one example that focuses on the *end* goal of development. In this model, how development is achieved is not necessarily as relevant as it facilitates the arrival of a status opposite to the existing one. Similarly, scholars argue that an overarching goal or "ultimate end" in the Sustainable Development Goals (SDGs) is critical for articulating the subgoals and targets and for providing synergies towards the larger goal (Costanza et al. 2014; Lim et al. 2018). Costanza et al. (2014) propose that "*equity & human well-being*—a prosperous, high quality of life that is equitably shared and sustainable" should be considered as the ultimate ends of sustainable development, while goals related to economic performance, technological advancements, political measures, and normative advocacies should be considered only as intermediate means of sustainable development. Such a view prioritises institutional change of the ends.[6]

Barriers to institutional change

If, for some reason, intended institutional changes did not occur, it is referred to as *institutional barriers* or *barriers of institutional change. Institutional barriers* in general are dynamic and context-dependent across sectoral, spatial, and temporal scales (Mimura et al. 2015). Literature has indicated three major types of *institutional barriers* to climate change adaptation in both developed and developing countries. The first type is barriers that impede *multi-level institutional coordination* between different political and administrative levels in society (Preston et al. 2013; Robinson and Berkes 2011). In climate change adaptation, some scholars have defended the need for top-down institutional change—national requirements or guidelines, regulations, policies, and codes—to occur, as currently climate adaptation is commonly inhibited by unclear roles and responsibilities across multiple levels and actors (Brown et al. 2013). The second type is the *institutional rigidity* which limits the horizontal interplay within local governments, where climate adaptation is seen as the isolated task of a singular sector (van den Berg and Coenen 2012; Runhaar et al. 2012; Wilby and Keenan 2012). In this case, past policies, decisions, habits, and traditions constrain the extent to which systems can learn or adapt to climate change, which scholars also refer to as *path dependency* (Preston et al. 2013; Runhaar et al. 2012). Sometimes *institutional rigidity* also exists in the form of *institutional barriers*, including competing values, conflicting objectives, tensions, and trade-offs in political dimensions, namely between different actors, policy agendas, and priorities in planning and implementation (Adger et al. 2009; Mimura et al. 2015; Storbjörk and Hedrén 2011). It is not uncommon for governments and local authorities to fail to present optimal solutions, possibly due to their cognitive and behavioural biases

[6] Williamson's levels of institutional change resonate with this broad categorisation of institutional change of the means and ends. Equity and human well-being are often considered as end goals of development. They are also a value or a norm, or a right which might not change frequently. If so, it would be a slow-changing institution at Williamson's embeddedness level.

(Krueger 1990; Podsakoff et al. 1990). The third type is barriers that impede *coordination between formal governmental and administrative agencies and social and private stakeholders* to create participatory approaches suitable for the localised and contextual nature of climate adaptation (Engle and Lemos 2010; Jonsson et al. 2012). Scholars consider private sector involvement to increase the efficiency of climate adaptation (Engle and Lemos 2010; Tompkins and Eakin 2012). Besides, key actors, advocates, and champions are decisive for initiating, mainstreaming, and sustaining momentum for climate adaptation planning and implementation in different national settings (Moser and Ekstrom 2010; Romero-lankao and Wilhelmi 2018).

3.5 Analysing Common-Pool Resource Governance Through the IAD Framework

Common-pool resource governance can be analysed through the Institutional Analysis and Development (IAD) framework (see Fig. 3.5). The IAD framework was intentionally developed to "map" out institutions and help people understand how institutions promote or obstruct human collaboration and governance (Ostrom 2005). Ostrom and her colleagues developed the IAD framework based on many empirical studies that revealed the complexity and diversity of the field settings (Kiser and Ostrom 2000; Ostrom 2005; Ostrom and Ostrom 1977). The meta-theoretical language enables researchers to discuss any theory or compare theories and identify influential variables in various empirical settings. The IAD framework has already been widely applied in empirical studies of CPRs such as irrigation systems, forest management, and greenhouse gas (GHG) emission for climate change mitigation, as well as within urban contexts (e.g., Mincey et al. 2013; Ostrom 1972; Ostrom et al. 1979).

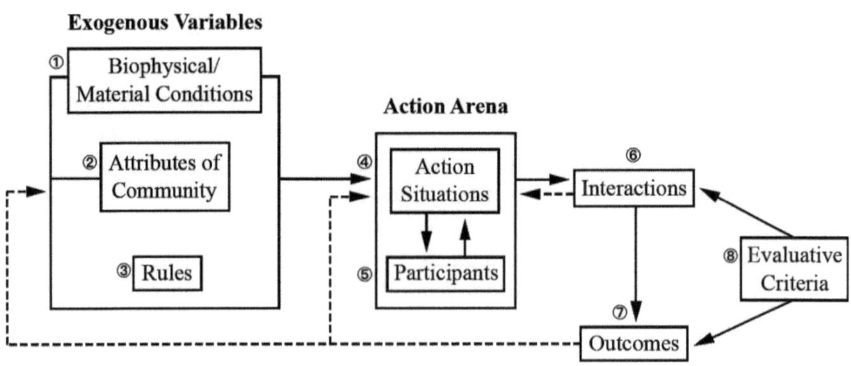

Fig. 3.5 Institutional analysis and development framework. *Source* Adapted from Ostrom (2005: 15)

Ostrom and her colleagues designed the IAD framework with a group of elements so that when analysing the governance of a particular resource, depending on the researcher's needs, one can analyse each of the elements multiple times. These elements are: ① *biophysical conditions*, ② *community attributes,* ③ *rules-in-use*, which can affect an ④ *action situation*, often considered the core of the IAD framework, ⑤ *actors or participants,* ⑥ *interactions,* ⑦ *outcomes*, and ⑧*evaluative criteria.*

① *Biophysical conditions* refer to the natural environmental conditions in which the institutional arrangements are applied. Sometimes it can be simplified as one of the four types of economic goods (public, private, club, and common-pool goods), as Ostrom defines (2005). It could also refer to general conditions such as climate change, environmental degradation, and resource scarcity in general.

② *Attributes of community* are the socio-economic as well as cultural, historical, and political characteristics of a community. They could include the history of prior interactions, internal homogeneity or heterogeneity of key attributes, and the knowledge and social capital of those who could participate or be affected by others.

③ *Rules-in-use* include regulations, norms, agreements, agendas, or anything that has a bidding effect—legal, political, or moral, to the actors involved in a decision-making setting potentially across several levels. They are the information which forms a common understanding of those involved related to who must, must not, or may take which actions affecting others depending on incentives or sanctions (Crawford and Ostrom 1995). Ostrom (2005) defines seven types of rules-in-use:

- *Boundary rules* define which actors are included based on specific criteria, e.g., geographical boundary, natural and socio-demographic conditions such as age and level of education.
- *Position rules* define in what way actors are participating in decision-making.
- *Choice rules* define what actors can do, must or must not do.
- *Information rules* define what information can be circulated among which actors, in what frequency and language.
- *Aggregation rules* define how actors collectively make decisions.
- *Scope rules* define the range of possible results within the action situation.
- *Payoff rules* define reward or sanction.

Ostrom further specifies that the ③ *Rules-in-use* may evolve or self-consciously change the rules in a collective-choice or constitutional-choice setting (Ostrom 2005), which could directly influence an ④ *Action situation.*

These elements together impact an ④ *Action situation*, which then generate patterns of interactions and outcomes. These outcomes may be evaluated by participants in the *action situation*, and potentially by researchers, which further feeds back on both the external variables and the *action situation* (Ostrom 2005). The ④ *Action situation* is often treated as the core part or the focal point of the IAD framework, as

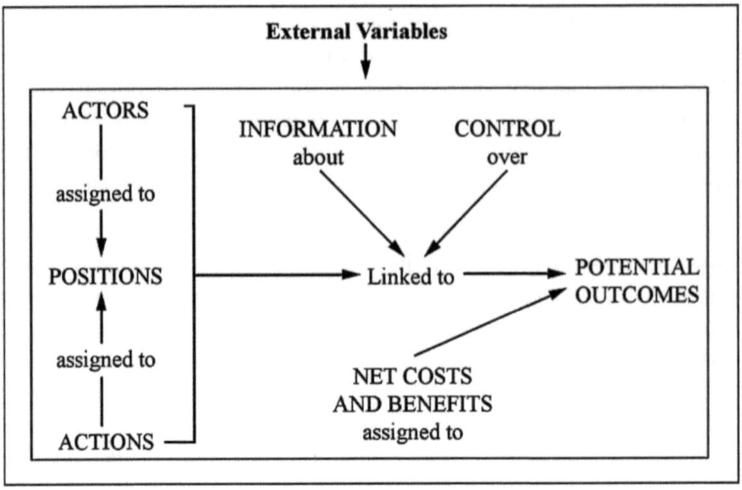

Fig. 3.6 Internal structure of an action situation. *Source* Adapted from Ostrom (2005: 33)

participants interact with each other within an *action situation* from the *position* they are assigned to, and make decisions depending on the *information*, power to *control*, and the *net costs and benefits* given. The internal structure of an *action situation* is demonstrated in Fig. 3.6. The costs and benefits include direct, monetary costs and benefits such as the costs of planting, pruning, removal, irrigation, and the benefits of cooling energy savings; as well as indirect, non-monetary costs and benefits, such as interception of particulates, stormwater runoff reduction, residents' relative well-being, and influences on local biodiversity (McPherson 1992).

⑤ *Actors or participants* of CPR governance, which may include governments and non-state actors.

⑥ *Interactions.* The original IAD framework automatically assumes that all types of actors within an action arena participate by interacting with other actors, such as agreeing, coordinating, cooperating, negotiating, or supervising. However, in this research, the researcher argues that not all situations fit this assumption. Hence, it is necessary to include the possibilities of non-participation, which also includes two possibilities: (1) not informed and not participating, and (2) informed but not participating. Only based on information can interactions happen. Based on information, two types of participation are anticipated: partial and substantial participation, in which actors agree, coordinate, cooperate, negotiate, or supervise (see Table 8: Categories, codes, and subcodes for the qualitative content analysis using MaxQDA as well as the subsequent analysis in Chap. 6).

The ⑦ *Outcomes* may include, (1) decisions and actions on a certain governance issue, such as Urban green spaces planning and implementation, and (2) the resultant formal and informal institutional changes, e.g., on the positions of actors and the property rights they hold.

⑧ *Evaluative criteria* will be established to evaluate ⑥ *Interactions* and ⑦ *Outcomes*. The original evaluative criteria defined by Ostrom (2005) are:

1. *Economic efficiency*. It is the flow of net benefits associated with an allocation or reallocation of resources.
2. *Equity*. It includes procedural equity and substantive equity, e.g., the right to participate is a procedural equity and the health and well-being benefits is a kind of substantive equity.
3. *Adaptability, resilience, and robustness*. It relates to the effect of intervention/implementation on the community, the environment, and the institutions overtime.
4. *Accountability*: It refers to the acknowledgement and assumption of responsibility by respective actors for their actions at the corresponding level or the next upper level.
5. *Conformance to general morality*. It refers to three aspects of morality—informational, legal, and ethical.

Now, despite that the IAD framework has been well acclaimed by scholars for its overall potential in analysing CPR governance, the ⑧ *evaluative criteria* is the least developed part subjected to future contestation as scholar research advances (Cole 2017). Considering that it is also the most flexible part of the IAD framework that allows context-specific readjustments, in this research, these five criteria are used as a structural reference, according to which a more comprehensive set of criteria is established (see Sect. 4.3.2.2 Applying multi-criteria assessment in this research).

Research on governance issues in China using the IAD framework is emerging quickly and seems to have successfully addressed a broad range of topics, particularly since Ostrom won the Nobel Prize in Economics in 2009. This emerging research primarily concentrates on water/irrigation system management (Araral and Wu 2016; Fan et al. 2019; Wang et al. 2017), conflicts of land-use/expropriation (Cao and Zhang 2018; Wu et al. 2018), urbanisation issues (Ma et al. 2019; Xiong et al. 2020), unexpected natural and human events (disasters and protests) (Liu and de Jong 2017; Ran et al. 2020). Among these studies, Ma et al. (2019) used the IAD framework to analyse Chinese farmers' willingness to settle in urban areas and its influencing factors. Fan et al. (2019) used the IAD framework to assess the effects of community perceptions and institutional capacity on smallholder farmers' responses to water scarcity in China's arid Northwestern region. Liu and de Jong (2017) used the institutional design principles of the IAD framework and the concept of common-pool resource to understand the institutional causes of environmental protests in China. Ran et al. (2020) studied geo-disasters governance in the Three Gorges Reservoir Region in China under the IAD framework. It is evident that scholars are increasingly finding the IAD framework convincing for analysing a vast range of action situations regardless of context. This is also thanks to the powerful meta-language that Ostrom (2005) and her colleagues were able to develop after decades of theoretical inquiry and empirical testing.

References

Acemoğlu D, Robinson J (2012) Why nations fail: the origins of power, prosperity, and poverty, 1st ed. Crown Business. https://www.goodreads.com/book/show/12158480-why-nations-fail

Adger WN, Dessai S, Goulden M, Hulme M, Lorenzoni I, Nelson DR, Naess LO, Wolf J, Wreford A (2009) Are there social limits to adaptation to climate change? Clim Change. https://doi.org/10.1007/s10584-008-9520-z

Alchian AA (1989) Property rights. In: Eatwell J, Milgate M, Newman P (eds) The invisible hand. Palgrave Macmillan, UK, pp 232–238. https://doi.org/10.1007/978-1-349-20313-0_31

Andonova LB, Betsill MM, Bulkeley H (2009) Transnational climate governance. Glob Environ Polit. https://doi.org/10.1162/glep.2009.9.2.52

Araral E, Wu X (2016) Comparing water resources management in China and India: policy design, institutional structure and governance. Water Policy. https://doi.org/10.2166/wp.2016.001

Armson D, Stringer P, Ennos AR (2012) The effect of tree shade and grass on surface and globe temperatures in an urban area. Urban Forest Urban Greening 11(3):245–255. https://doi.org/10.1016/j.ufug.2012.05.002

Arrow KJ, Kenneth J (1974) The limits of organization. Norton. https://books.google.com.sg/books/about/The_Limits_of_Organization.html?id=55eDQgAACAAJ&redir_esc=y

van den Berg M, Coenen F (2012) Integrating climate change adaptation into Dutch local policies and the role of contextual factors. Local Environ. https://doi.org/10.1080/13549839.2012.678313

Berkes F (2006) From community-based resource management to complex systems: the scale issue and marine commons. Ecol Soc. https://doi.org/10.5751/ES-01431-110145

Berkes F (2007) Community-based conservation in a globalized world. Proc Nat Acad Sci. https://doi.org/10.1073/pnas.0702098104

Biermann F, Betsill MM, Gupta J, Kanie N, Lebel L, Liverman D, Schroeder H, Siebenhüner B, Zondervan R (2010) Earth system governance: a research framework. Int Environ Agreements: Polit Law Econ. https://doi.org/10.1007/s10784-010-9137-3

Blomquist W, Ostrom E (2008) Deliberation, learning, and institutional change: the evolution of institutions in judicial settings. Constit Polit Econ. https://doi.org/10.1007/s10602-008-9045-5

Blomquist W (2009) Multi-level governance and natural resource management: the challenges of complexity, diversity, and uncertainty. In: Institutions and sustainability: political economy of agriculture and the environment-essays in honour of Konrad Hagedorn. https://doi.org/10.1007/978-1-4020-9690-7_6

Bowen HR (1943) The interpretation of voting in the allocation of economic resources. Q J Econ 58(1):27–48. https://doi.org/10.2307/1885754

Bridges A (2016) The role of institutions in sustainable urban governance. Nat Resour Forum. https://doi.org/10.1111/1477-8947.12116

Bromley D (1986) The common property challenge. In: National Research Council (ed) In: Proceedings of the conference on common property resource management, Annapolis Maryland. National Academy Press, pp 1–6

Brown HCP, Smit B, Somorin OA, Sonwa DJ, Ngana F (2013) Institutional perceptions, adaptive capacity and climate change response in a post-conflict country: a case study from Central African Republic. Clim Dev. https://doi.org/10.1080/17565529.2013.812954

Cao Y, Zhang X (2018) Are they satisfied with land taking? Aspects on procedural fairness, monetary compensation and behavioral simulation in China's land expropriation story. Land Use Policy. https://doi.org/10.1016/j.landusepol.2017.08.027

Cash DW, Adger WN, Berkes F, Garden P, Lebel L, Olsson P, Pritchard L, Young O (2006) Scale and cross-scale dynamics: governance and information in a multilevel world. Ecol Soc. https://doi.org/10.5751/es-01759-110208

Castán Broto V (2017) Urban governance and the politics of climate change. In: World Dev. https://doi.org/10.1016/j.worlddev.2016.12.031

Chambwera M, Heal G, Dubeux C, Hallegatte S, Leclerc L, Markandya A, McCarl BA, Mechler R, Neumann JE, Calvo E, Iglesias A, Navrud S, Kairiza T (2015) Economics of adaptation. In:

Climate change 2014 impacts, adaptation and vulnerability: Part A: global and sectoral aspects. https://doi.org/10.1017/CBO9781107415379.022
Cheung SNS (1989) Economic organization and transaction costs. In: Eatwell J, Milgate M, Newman P (eds) Allocation, information and markets. Palgrave Macmillan UK, pp 77–82. https://doi.org/10.1007/978-1-349-20215-7_8
Chhatre A, Agrawal A (2009) Trade-offs and synergies between carbon storage and livelihood benefits from forest commons. Proc Natl Acad Sci. https://doi.org/10.1073/pnas.0905308106
Cimato F, Mullan M (2010) Defra evidence and analysis series—Adapting to climate change: analysing the role of government. http://www.nationalarchives.gov.uk/doc/
Coase RH (1937) The nature of the firm. Economica. https://doi.org/10.1111/j.1468-0335.1937.tb00002.x
Coase RH (1960) The problem of social cost. J Law Econ 3(October):1–44. https://www.law.uchicago.edu/files/file/coase-problem.pdf
Coase RH (1992) The institutional structure of production. Am Econ Rev. https://doi.org/10.1007/978-3-540-69305-5_3
Coase R (1998) The new institutional economics. Am Econ Rev 88(2):72–74. http://www.jstor.org/stable/116895
Coggan A, Whitten SM, Bennett J (2010) Influences of transaction costs in environmental policy. Ecol Econ. https://doi.org/10.1016/j.ecolecon.2010.04.015
Colding J, Barthel S, Bendt P, Snep R, van der Knaap W, Ernstson H (2013) Urban green commons: Insights on urban common property systems. Glob Environ Change. https://doi.org/10.1016/j.gloenvcha.2013.05.006
Cole DH (2017) Laws, norms, and the institutional analysis and development framework. J Inst Econ. https://doi.org/10.1017/S1744137417000030
Coleman EA, Steed BC (2009) Monitoring and sanctioning in the commons: an application to forestry. Ecol Econ. https://doi.org/10.1016/j.ecolecon.2009.02.006
Commission on Global Governance (1995) Our global neighborhood : the report of the commission on global governance, 1st edn. Oxford University Press
Costanza R, McGlade J, Lovins H, Kubiszewski I, Director I (2014) An overarching goal for the UN sustainable development goals. Solutions.
Crawford SES, Ostrom E (1995) A grammar of institutions. Am Polit Sci Rev. https://doi.org/10.2307/2082975
Degnbol P, McCay BJ (2007) Unintended and perverse consequences of ignoring linkages in fisheries systems. ICES J Marine Sci. https://doi.org/10.1093/icesjms/fsm040
Demsetz H (1974) Toward a theory of property rights. In: Classic papers in natural resource economics. https://doi.org/10.1057/9780230523210_9
Elmqvist T, Gatzweiler F, Lindgren E, Liu J (2019) Resilience management for healthy cities in a changing climate. In: Biodiversity and health in the face of climate change. Springer, pp 411–424. https://doi.org/10.1007/978-3-030-02318-8_18
Engle NL, Lemos MC (2010) Unpacking governance: building adaptive capacity to climate change of river basins in Brazil. Global Environ Change. https://doi.org/10.1016/j.gloenvcha.2009.07.001
Epstein G, Pittman J, Alexander SM, Berdej S, Dyck T, Kreitmair U, Raithwell KJ, Villamayor-Tomas S, Vogt J, Armitage D (2015) Institutional fit and the sustainability of social-ecological systems. Curr Opin Environ Sustain. https://doi.org/10.1016/j.cosust.2015.03.005
European Commission (2013) Green infrastructure (GI)—Enhancing Europe's natural capital—COM 149. Available at https://www.eea.europa.eu/themes/sustainability-transitions/urban-environment/urban-green-infrastructure/what-is-green-infrastructure
Fan Y, Tang Z, Park SC (2019) Effects of community perceptions and institutional capacity on smallholder farmers' responses to water scarcity: evidence from arid northwestern China. Sustainability (Switzerland). https://doi.org/10.3390/su11020483
Fankhauser S, Smith JB, Tol RSJ (1999) Weathering climate change: some simple rules to guide adaptation decisions. Ecol Econ 30(1):67–78. https://doi.org/10.1016/S0921-8009(98)00117-7

Frank B (2017) Urban systems: a socio-ecological system perspective. Soc Int J. https://doi.org/10.15406/sij.2017.01.00001
Gatzweiler FW (2003) The changing nature of economic value : indigenous forest garden values in Kalimanatan [sic] Indonesia. Shaker Verlag
Gibson CC, Ostrom E, Ahn TK (2000) The concept of scale and the human dimensions of global change: a survey. Ecol Econ. https://doi.org/10.1016/S0921-8009(99)00092-0
Gibson CC, Williams JT, Ostrom E (2005) Local enforcement and better forests. World Dev. https://doi.org/10.1016/j.worlddev.2004.07.013
Glaas E, Juhola S (2013) New levels of climate adaptation policy: analyzing the institutional interplay in the baltic sea region. Sustainability (Switzerland). https://doi.org/10.3390/su5010256
Grant M (2020) In: Stone V (ed) Integrating health in urban and territorial planning: A sourcebook, 1st edn. UN-Habitat and World Health Organization. https://unhabitat.org/sites/default/files/2020/05/1-final_highres_20002_integrating_health_in_urban_and_territorial_planning_a_sourcebook.pdf
Gregory McPherson E (1992) Accounting for benefits and costs of urban greenspace. Landscape Urban Plan. https://doi.org/10.1016/0169-2046(92)90006-L
Hagedorn K (2008) Integrative and segregative institutions: a dichotomy for understanding institutions of sustainability. In: Schäfer C, Rupschus C, Nagel U-J (eds) Enhancing the capacities of agricultural systems and producers, proceedings of the second green week scientific conference. Margraf-Publishers. www.agrar.hu-berlin.de/wisola/fg/ress/
Hagedorn K (2015) Can the concept of integrative and segregative institutions contribute to the framing of institutions of sustainability? Sustainability (Switzerland). https://doi.org/10.3390/su7010584
Hartig T, Mitchell R, de Vries S, Frumkin H (2014) Nature and health. Annu Rev Public Health 35(1):207–228. https://doi.org/10.1146/annurev-publhealth-032013-182443
Heltberg R (2002) Property rights and natural resource management in developing countries. J Econ Surv 16(2):189–214. http://www.gsdrc.org/docs/open/ssaj39.pdf
Ho P (2017) The "Credibility book" and its application to property rights. In: Unmaking China's development. https://doi.org/10.1017/9781316145616.005
Hodgson GM (2006) What are institutions? J Econ Issues. https://doi.org/10.1080/00213624.2006.11506879
ICLEI—Local Governments for Sustainability (2017) Nature-based solutions for sustainable urban development. https://unfccc.int/files/parties_observers/submissions_from_observers/application/pdf/777.pdf
IPCC (2014) Summary for policymakers. Climate change 2014: impacts, adaptation and vulnerability—Contributions of the working group II to the fifth assessment report, pp 1–32. https://doi.org/10.1016/j.renene.2009.11.012
Jia H, Yu S, Qin H (2017) Low impact development and sponge city construction for urban stormwater management. Front Environ Sci Eng 11(4). https://doi.org/10.1007/s11783-017-0989-4
Jonsson AC, Hjerpe M, Andersson-Sköld Y, Glaas E, André K, Simonsson L (2012) Cities' capacity to manage climate vulnerability: experiences from participatory vulnerability assessments in the lower Göta Älv Catchment, Sweden. Local Environ. https://doi.org/10.1080/13549839.2012.685880
Kelly DL, Kolstad CD, Mitchell G (2005) Adjustment costs from environmental change, J Environ Econ Manage, 50(3):468–495
Kiser LL, Ostrom E (2000) The three worlds of action: a metatheoretical synthesis of institutional approaches. In: Polycentric games and institutions: readings from the workshop in political theory and policy analysis
Korten DC (1987) Third generation NGO strategies: a key to people-centered development. World Dev. https://doi.org/10.1016/0305-750X(87)90153-7

Kostka G (2015) Command without control: the case of China's environmental target system. SSRN Electron J. https://doi.org/10.2139/ssrn.2613427

Krueger AO. (1990). Government failures in development. J Econ Perspect 4(3):9–23. https://pubs.aeaweb.org/doi/pdf/10.1257/jep.4.3.9

Kuzevanov VY, Gubiy EV (2014) Botanic gardens as world ecological resources for innovative technological development. In: Proceedings of Irkutsk State University 10:73–81. http://bogard.isu.ru/articles/2010_tomsk/kuz_tomsk.pdf

Larondelle N, Haase D, Kabisch N (2014) Mapping the diversity of regulating ecosystem services in European cities. Glob Environ Change. https://doi.org/10.1016/j.gloenvcha.2014.04.008

Latour B, Douglas M (1988) How institutions think. Contemp Sociol. https://doi.org/10.2307/2069673

Li Y, Shi Y, Qureshi S, Bruns A, Zhu X (2014) Applying the concept of spatial resilience to socio-ecological systems in the urban wetland interface. Ecol Indicators. https://doi.org/10.1016/j.ecolind.2013.09.032

Libecap GD (2014) Addressing global environmental externalities: transaction costs considerations. J Econ Liter. https://doi.org/10.1257/jel.52.2.424

Lim MML, Søgaard Jørgensen P, Wyborn CA (2018) Reframing the sustainable development goals to achieve sustainable development in the anthropocene—a systems approach. Ecol Soc. https://doi.org/10.5751/ES-10182-230322

Liu J, Gatzweiler FW (2020) The institutional challenge to co-deliver migrant integration and urban greening—evidence from Haizhu Wetland Park Project in Guangzhou, China. J Chin Govern. https://doi.org/10.1080/23812346.2020.1760069

Liu L, de Jong M (2017) The institutional causes of environmental protests in China: a perspective from common pool resource management. J Chin Govern. https://doi.org/10.1080/23812346.2017.1354432

Lyytimäki J (2013) Nature's nocturnal services: Light pollution as a non-recognised challenge for ecosystem services research and management. Ecosyst Serv 3:e44–e48. https://doi.org/10.1016/j.ecoser.2012.12.001

Ma L, Chen M, Che X, Fang F (2019) Farmers' rural-to-urban migration, influencing factors and development framework: a case study of sihe village of Gansu, China. Int J Environ Res Public Health. https://doi.org/10.3390/ijerph16050877

McCann L, Colby B, Easter KW, Kasterine A, Kuperan KV (2005) Transaction cost measurement for evaluating environmental policies. Ecol Econ. https://doi.org/10.1016/j.ecolecon.2004.08.002

McPhearson T, Haase D, Kabisch N, Gren Å (2016) Advancing understanding of the complex nature of urban systems. Ecol Indicators. https://doi.org/10.1016/j.ecolind.2016.03.054

McPhearson T, Pickett STA, Grimm NB, Niemelä J, Alberti M, Elmqvist T, Weber C, Haase D, Breuste J, Qureshi S (2016) Advancing urban ecology toward a science of cities. BioScience. https://doi.org/10.1093/biosci/biw002

Meinzen-Dick R (2007) Beyond panaceas in water institutions. Proc Nat Acad Sci. https://doi.org/10.1073/pnas.0702296104

Mimura N, Pulwarty RS, Duc DM, Elshinnawy I, Redsteer MH, Huang HQ, Nkem JN, Rodriguez RAS, Moss R, Vergara W, Darby LS, Kato S (2015) Adaptation planning and implementation. In: Climate change 2014 impacts, adaptation and vulnerability: Part A: global and sectoral aspects. https://doi.org/10.1017/CBO9781107415379.020

Mincey SK, Hutten M, Fischer BC, Evans TP, Stewart SI, Vogt JM (2013) Structuring institutional analysis for urban ecosystems: a key to sustainable urban forest management. Urban Ecosyst 16(3):553–571. https://doi.org/10.1007/s11252-013-0286-3

Moser SC, Ekstrom JA (2010) A framework to diagnose barriers to climate change adaptation. Proc Natl Acad Sci. https://doi.org/10.1073/pnas.1007887107

Musgrave RA (1939) The voluntary exchange theory of public economy. Q J Econ 53(2):213–237. https://doi.org/10.2307/1882886

Nagendra H, Ostrom E (2012) Polycentric governance of multifunctional forested landscapes. Int J Commons. https://doi.org/10.18352/ijc.321

Natural Resource Governance Institute (2017) Natural resource governance glossary, p 14. https://resourcegovernance.org/sites/default/files/documents/natural-resource-governance-glossary.pdf
Neale WC (1993) Institutions, institutional change and economic performance. Douglass C. North. Econ Dev Cultural Change. https://doi.org/10.1086/452020
Nesshöver C, Assmuth T, Irvine KN, Rusch GM, Waylen KA, Delbaere B, Haase D, Jones-Walters L, Keune H, Kovacs E, Krauze K, Külvik M, Rey F, van Dijk J, Vistad OI, Wilkinson ME, Wittmer H (2017) The science, policy and practice of nature-based solutions: an interdisciplinary perspective. Sci Total Environ. https://doi.org/10.1016/j.scitotenv.2016.11.106
Newig J, Fritsch O (2009) Environmental governance: participatory, multi-level—And effective? Environ Policy Govern. https://doi.org/10.1002/eet.509
Nordhaus WD (2009) Paul Samuelson and global public goods. In: Samuelsonian economics and the twenty-first century. https://doi.org/10.1093/acprof:oso/9780199298839.003.0006
North DC (1984) Transaction costs, institutions, and economic history. J Inst Theor Econ
North DC (1990) A transaction cost theory of politics. J Theor Polit. https://doi.org/10.1177/0951692890002004001
North DC (1991) Institutions. J Econ Perspect 5(1):97–112. https://pubs.aeaweb.org/doi/pdf/10.1257%2Fjep.5.1.97
North DC, North DC (2012) An introduction to institutions and institutional change. In: Institutions, institutional change and economic performance. https://doi.org/10.1017/cbo9780511808678.003
Olsen E, Kleiven AR, Skjoldal HR, von Quillfeldt CH (2011) Place-based management at different spatial scales. J Coastal Conserv. https://doi.org/10.1007/s11852-010-0108-1
Ostrom E, Parks RB, Whitaker G (1979) Patterns of metropolitan policing. Am Polit Sci Rev 73(2):596–597. https://doi.org/10.2307/1954941
Ostrom E, Nagendra H (2006) Insights on linking forests, trees, and people from the air, on the ground, and in the laboratory. Proc Natl Acad Sci. https://doi.org/10.1073/pnas.0607962103
Ostrom V, Ostrom E (1977) Public goods and public choices. https://doi.org/10.1007/978-3-662-44143-5_5
Ostrom V, Tiebout CM, Warren R (1961) The organization of government in metropolitan areas: a theoretical inquiry. Am Econ Rev. https://doi.org/10.1017/S0003055400125973
Ostrom E (1972) Metropolitan reform: propositions derived from two traditions. Soc Sc Quart
Ostrom E (1990) Governing the commons—The evolution of institutions for collective action, 1st edn. Cambridge University Press. http://www.ambridge.org
Ostrom E (2005) Understanding institutional diversity. In: Princeton University Press
Ostrom E (2010) Beyond markets and states: polycentric governance of complex economic systems. Am Econ Rev. https://doi.org/10.1257/aer.100.3.641
Ostrom E (2015) Governing the commons: the evolution of institutions for collective action. In: Governing the commons: the evolution of institutions for collective action. Cambridge University Press. https://doi.org/10.1017/CBO9781316423936
Paavola J, Adger WN (2005) Institutional ecological economics. Ecol Econ. https://doi.org/10.1016/j.ecolecon.2004.09.017
Panno A, Carrus G, Lafortezza R, Mariani L, Sanesi G (2017) Nature-based solutions to promote human resilience and wellbeing in cities during increasingly hot summers. Environ Res. https://doi.org/10.1016/j.envres.2017.08.016
Podsakoff PM, MacKenzie SB, Moorman RH, Fetter R (1990) Transformational leader behaviors and their effects on followers' trust in leader, satisfaction, and organizational citizenship behaviors. Leadership Quart. https://doi.org/10.1016/1048-9843(90)90009-7
Preston BL, Dow K, Berkhout F (2013) The climate adaptation frontier. Sustainability (Switzerland). https://doi.org/10.3390/su5031011
Ran A, Fan J, Zhou L, Zhang C (2020) Geo-disaster governance under the IAD framework: the case study of Chongqing's three Gorges Reservoir Region, China. Sustainability (Switzerland). https://doi.org/10.3390/su12145517

Revi A, Satterthwaite DE, Aragón-Durand F, Corfee-Morlot J, Kiunsi R, Pelling M, Roberts D, Solecki W (2014) 8. Urban areas. In: Climate change 2014: impacts, adaptation, and vulnerability. Part A: global and sectoral aspects. contribution of working group II to the fifth assessment report of the intergovernmental panel on climate change. https://doi.org/10.1017/CBO9781107415379.013

Robertson J (1994) People-centered development: principles for a new civilization. People-Centered Development Forum. https://davidkorten.org/07robert/

Robinson LW, Berkes F (2011) Multi-level participation for building adaptive capacity: formal agency-community interactions in northern Kenya. Glob Environ Change. https://doi.org/10.1016/j.gloenvcha.2011.07.012

Rodima-Taylor D, Olwig MF, Chhetri N (2012) Adaptation as innovation, innovation as adaptation: an institutional approach to climate change. Appl Geogr. https://doi.org/10.1016/j.apgeog.2011.10.011

Romero-lankao P, Wilhelmi O (2018) Chapter 4 : live with risk while reducing vulnerability. May, pp 92–112

Rothman J (2019) The art of decision-making. The New Yorker. https://www.newyorker.com/magazine/2019/01/21/the-art-of-decision-making

Runhaar H, Mees H, Wardekker A, van der Sluijs J, Driessen PPJ (2012) Adaptation to climate change-related risks in Dutch urban areas: stimuli and barriers. Regional Environ Change. https://doi.org/10.1007/s10113-012-0292-7

Samuelson PA, Nordhaus WD (2010) Microeconomics, 19th edn. McGraw-Hill/Irwin

Samuelson PA (1954) The pure theory of public expenditure. Rev Econ Stat 36(4):387–389. https://courses.cit.cornell.edu/econ335/out/samuelson_pure.pdf

Samuelson PA (1955) Diagrammatic exposition of a theory of public expenditure. Rev Econ Stat 37(4):350–356. http://www.socsci.uci.edu/~jkbrueck/coursereadings/Econ272Breadings/samuelson.pdf

Schlager E, Heikkila T (2011) Left high and dry? climate change, common-pool resource theory, and the adaptability of western water compacts. Publ Admin Rev. https://doi.org/10.1111/j.1540-6210.2011.02367.x

Schlager E (2004) Common-pool resource theory. In: Environmental governance reconsidered: challenges, choices, and opportunities. MIT Press, pp 145–175. https://dlc.dlib.indiana.edu/dlc/bitstream/handle/10535/5648/Common-poolResourceTheory.pdf?sequence=5

Simon HA (1956) Rational choice and the structure of the environment. Psychol Rev 63(2):129–138. https://pdfs.semanticscholar.org/23a9/4ce42fe0d50f5c993f34d4c9602f8aeac507.pdf

Simon HA (1990) Bounded rationality. In: Eatwell J (ed) Utility and probability. Palgrave Macmillan, UK, pp 15–18. https://doi.org/10.1007/978-1-349-20568-4_5

Storbjörk S, Hedrén J (2011) Institutional capacity-building for targeting sea-level rise in the climate adaptation of Swedish coastal zone management. Lessons from Coastby. Ocean Coastal Manage. https://doi.org/10.1016/j.ocecoaman.2010.12.007

Tompkins EL, Eakin H (2012) Managing private and public adaptation to climate change. Global Environ Change. https://doi.org/10.1016/j.gloenvcha.2011.09.010

Trawick PB (2001) Successfully governing the commons: Principles of social organization in an Andean irrigation system. Hum Ecol. https://doi.org/10.1023/A:1007199304395

Trenberth KE (2008) Observational needs for climate prediction and adaptation. WMO Bull

Tyler S, Moench M (2012) A framework for urban climate resilience. Clim Dev. https://doi.org/10.1080/17565529.2012.745389

Vatiero M (2020) The theory of transaction in institutional economics: A history, 1st edn. Routledge. https://www.routledge.com/The-Theory-of-Transaction-in-Institutional-Economics-A-History/Vatiero/p/book/9780367194697

Vatn A (2005) Institutions and the environment. Edward Elgar Publications

Wang RY, Ng CN, Lenzer JH, Dang H, Liu T, Yao S (2017) Unpacking water conflicts: a reinterpretation of coordination problems in China's water-governance system. Int J Water Resour Dev. https://doi.org/10.1080/07900627.2016.1197824

Wang Y (2017) Enhancing commons governance: elinor ostrom's scholarship and applications (In Chinese: 增进公共事务治理-奥斯特罗姆学术探微与应用). Tsinghua University Press. http://www.bookschina.com/7479709.htm

Wegerich K (2001) Institutional change: a theoretical approach. https://www.soas.ac.uk/water/publications/papers/file38373.pdf

Welden EA, Chausson A, Melanidis MS (2021) Leveraging nature-based solutions for transformation: reconnecting people and nature. People Nat 00:1–12. https://doi.org/10.1002/PAN3.10212

Wilby RL, Keenan R (2012) Adapting to flood risk under climate change. Progr Phys Geogr. https://doi.org/10.1177/0309133312438908

Williamson OE (1977) Markets and hierarchies. Challange

Williamson OE (1979) The University of Chicago The Booth School of Business of the University of Chicago The University of Chicago Law School Transaction-Cost Economics : The governance of contractual relations transaction-cost economics : the governance of contractual relation. J Law Econ1 22(2):233–261

World Bank (2017) World development report 2017: governance and the law. http://www.worldbank.org/en/publication/wdr2017

World Health Organization, WHO, World Health Organization, WHO, World Health Organization, WHO, World Health Organization, WHO, World Health Organization, & WHO (2015).WHO global strategy on people-centred and integrated health services. In: WHO/HIS/SDS/2015.6

Wu Y, Mo Z, Peng Y, Skitmore M (2018) Market-driven land nationalization in China: a new system for the capitalization of rural homesteads. Land Use Policy. https://doi.org/10.1016/j.landusepol.2017.10.003

Xiong W, Chen B, Wang H, Zhu D (2020) Public–private partnerships as a governance response to sustainable urbanization: lessons from China. Habitat Int. https://doi.org/10.1016/j.habitatint.2019.102095

Young O (2002) The institutional dimensions of environmental change: fit, interplay, and scale. Organization. https://doi.org/10.1023/A:1024850005580

Zhao Y (2005) Urban planning in the perspective of institutional economics. City Plan Rev 29(06&07):40–47, 17–27. http://en.cnki.com.cn/Article_en/CJFDTOTAL-CSGH200506008.htm

Zölch T, Henze L, Keilholz P, Pauleit S (2017) Regulating urban surface runoff through nature-based solutions—An assessment at the micro-scale. Environ Res. https://doi.org/10.1016/j.envres.2017.05.023

Chapter 4
Research Design and Methodology

4.1 Research Design

4.1.1 Designing the Empirical Analysis with the IAD Framework

The previous chapter established the IAD framework as an analytical tool for dissecting and examining the quality of CPR governance cases in empirical settings. This research applies the IAD framework to understand three Urban green spaces governance cases in Guangzhou, China. It is used in three ways:

1. define independent and dependent variables
2. confirm this research's core variables, which are the ③ *rule-in-use* and the ④ *action situation*
3. guide the composition of the in-depth and semi-structured interview questions, aiming to find adequate and consistent information about the different variables in three case studies.

First, the IAD framework is used here to define the independent and dependent variables. The three exogenous variables: ① *biophysical conditions*, ② *community attributes,* and ③ *rules-in-use* are the independent variables at the beginning of the research time frame. Within the research time frame, they gradually become dependent variables as the ⑦ *outcomes* feedback to them (see Fig. 4.1). The ① *biophysical conditions* in this research refer to the conditions of the Urban green spaces in the city of Guangzhou, including wetlands, urban parks, and ecological corridors. For ② *community attributes*, three community clusters—local governments, communities or residents, and landowners, each has their attributes. Local governments include the city and district governments, communities, or residents are those who reside and/or make their living within the study area. Landowners refer to those who legally own the land and/or the property above who do not necessarily reside or make their living

© Zhejiang University Press 2023

J. Liu, *Transforming Urban Green Space Governance in China Under Ecological Civilization: An Institutional Analysis*, Urban Health and Wellbeing,
https://doi.org/10.1007/978-981-99-6694-3_4

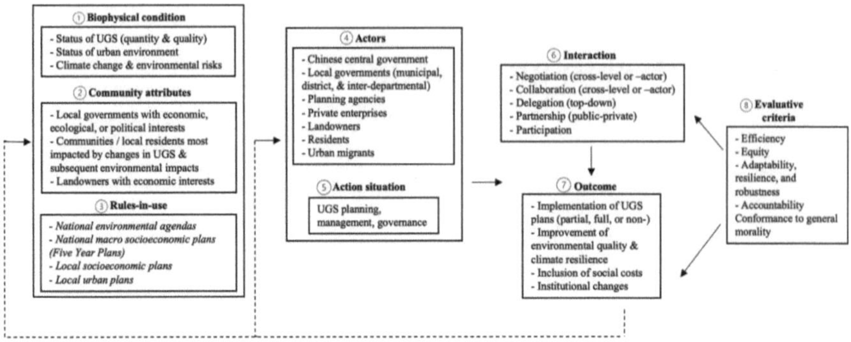

Fig. 4.1 IAD framework applied in this research

within the study area. The ③ *rules-in-use* include the national development agendas, urban macro-plans, local norms, or regional preferences of urban development, as well as the legal and political positions where various public and private actors across several levels are placed in decision-making settings regarding Urban green spaces planning and management in the city of Guangzhou.

Second, the IAD framework is used to confirm the core variables of this research, which are the ③ *rule-in-use* and the ④ *action situation*. Particularly, ④ *action situation* is the focal point where various actors communicate, negotiate, agree, or disagree, collaborate or delegate towards certain outcomes in the three Urban green spaces planning and governance cases, including the implementation of Urban green spaces plans, the improvement of environmental quality and climate resilience, the inclusion of social costs, and potential institutional changes.

Furthermore, the IAD framework is also used to structure the in-depth and semi-structured interview questions, aiming to find adequate and consistent information about the different variables in three case studies. Table 4.1 shows the four main research questions and their correspondent variables and reveals the methods of data collection and analysis suitable for analysing the variables to answer these main research questions.

4.1.2 A Qualitative, Case Study-Based Research Approach

The institutional focus of this research determines its qualitative and case-based approach. Scholars have applied institutional analysis on environmental and social challenges as long-term consequences of formal and informal institutional arrangements interplayed at different levels and scales, most of which were qualitative studies (Cumming et al. 2006; Young 2002). The complexity of real-world decision-making contexts and processes means that many factors or variables can be too intricate to be accurately and comprehensively described by quantitative methods (Schlüter 2010).

Table 4.1 Overview of research design

Aims	Research questions	Key variables	Data collection methods	Data analysis methods
A1. Understand institutional dynamics, facilitators, and underlying barriers	Q1. What institutional facilitations or challenges have Chinese urban governments received or confronted when planning and managing Urban green spaces?	① Biophysical condition ② Community attributes ③ Rule-in-use ⑤ Actors ⑦ Outcome ⑧ Evaluative criteria	In-depth interviews Archive research	Qualitative content analysis
A2. Examine the extent and approaches for Urban green spaces upscaling	Q2. How have local governments managed to upscale Urban green spaces towards more effective Nature-based Solutions?	① Biophysical condition ③ Rule-in-use ④ Action situation ⑤ Actors ⑥ Interactions ⑦ Outcome ⑧ Evaluative criteria	In-depth interviews Archive research	Qualitative content analysis Multi-criteria assessment (MCA)
A3. Analyse costs and benefits taken into consideration	Q3. What are the good practices to be shared?	③ Rule-in-use ⑥ Interactions ⑦ Outcomes ⑧ Evaluative criteria	In-depth interviews Archive research Observation	Qualitative content analysis MCA
A4. Understand intrinsic-value-related factors of institutional barriers	Q4. Why some challenges persisted despite centralised government system and strong political determination?	② Community attributes ③ Rule-in-use ⑤ Actors ⑥ Interactions ⑧ Evaluative criteria	In-depth interviews Archive research Observation	Qualitative content analysis MCA

A qualitative case study approach can carry out an in-depth description and analysis of decision-making contexts and processes and their complex scenarios without omitting factors caused by the assumption of preconditions. Ostrom's IAD framework is based on a meta-analysis of many case studies. The qualitative approach focuses on making sense of or interpreting social phenomena, e.g., experiences, behaviours, organisational functioning, social movements, interactional relationships, and cultural descriptions. The qualitative approach is also used to study phenomena in their natural settings instead of labs or experimental settings. Furthermore, a qualitative method allows researchers to interpret phenomena regarding the meanings actors bring to them (Denzin and Lincoln 2000).

This research aims to contextualise, interpret, and understand how different actors behave about Urban green spaces governance. The qualitative case study approach used in this research is an inductive and open process in which the researcher is the primary instrument for data collection and analysis through iteratively conducting fieldwork. The qualitative case study approach attends to descriptive data of low inference, focusing on the detailed description of *How* versus *How Many*. The research subject is context-specific in socio-economic, political, temporal, and environmental dimensions. Hence, the qualitative approach can better attend to the specificities of cases and their details through the researcher's contact with people and their insights. "The devil is in the detail." as the idiom goes, the qualitative approach can contribute to this research more than the quantitative.

4.1.3 Case Selection, Study Period, and Administrative Level

This research includes three cases of Urban green spaces planning and governance in the city of Guangzhou, all of which were implemented within the last two decades: an ecological corridor, an urban wetland park converted from an orchard, and a Sponge City demonstration site (for the definition of *Sponge City* please see Sect. 5.3.4 *Tianhe Sponge City Demonstration Site*). Choosing all three case studies in the same city allows this research to explore the interaction of multiple actors, various organisations, and the complex arrangements of each case within the same institutional, geographical, natural, cultural, socio-economic, and political context. Guangzhou is an ideal city for studying Urban green spaces planning and governance due to the city's natural and urban condition: an emerging, compact megacity located in a low-lying estuary and in a subtropical climate zone with a high scarcity in land resource and high risk of climate change impacts (see Sect. *1.1 Urbanisation Trend, Environmental Degradation, and Climate Change Impacts in* China). The natural, socio-economic context of Guangzhou will be introduced in greater detail in Sect. 5.1.1 Natural and Socio-economic Conditions.

As for the study period, this research defines the study period of empirical cases between the New Millennium and the current time for two main reasons. All three chosen case studies incurred since 2000. Firstly, in the New Millennium, for the first time, the Chinese central government included *sustainable development* as one of the primary goals in the nation's *10th Five-Year Plan (FYP) for Socio-Economic Development*, which states "the main expected goals of sustainable development are: … The trend of ecological deterioration will be curbed, the forest coverage rate will be increased to 18.2%, and the green coverage rate in urban built areas will be increased to 35%. The urban and rural environmental quality will be improved, and the total discharge of major pollutants will be decreased by 10% compared with 2000" (State Council of China 2001). Secondly, Guangzhou's Municipal Government considered the New Millennium as a great timing for long-term strategic urban development planning to enable and sustain further urbanisation. A macro-level strategic urban development plan was created, which included the merging of two neighbouring counties—Panyu County and Huadu—into the city's metropolitan territory, and a conceptual plan of an ecological corridor network (also termed the "Ecological City" plan) throughout the new metropolitan area (Guangzhou Daily 2018).

The ecological corridor network resulting from the conceptual plan became this research's first case study of Urban green space planning and governance. The second case—the Haizhu Wetland Park—is another major case of Urban green spaces planning and governance in Guangzhou that occurred around 2011, a decade after the first case—the ecological corridor network conceptual plan. The second case is relevant as it explored a new urban green space planning and governance model by transferring property rights. The third case—the Tianhe Sponge City demonstration site—was implemented around 2015 with the specific goal of adapting to the increasing urban inundation and flash floods aggravated by climate change. The third case represents a new kind of urban green space planning and governance for the technical and political challenges it presented.

On the administrative level, this research focuses on urban green space planning and governance in China at the city/municipal level for two main reasons. First, Chinese cities, as the local administrative units, make independent plans and policies on Urban green spaces governance depending on their status of land-use and economic and social needs based on the broad policy guidance provided or the goals proposed in the national Five-Year Plans (FYPs) by the central government. Second, the local level directly absorbs the externalities induced by improper urban green space planning and governance, prompting the need to focus on the local level. More about the central-local decision-making and coordination mechanism in the Chinese administrative system can be found in Sect. 5.2.4 A Multi-level, Nested Hierarchical System of Land-Use and Spatial Planning.

4.2 Data Collection

This research uses Grounded Theory as a methodological philosophy for data collection and analysis. As the term suggests, Grounded Theory is a general, inductive methodology "grounded" in qualitative data that is systematically gathered and analysed (Charmaz 2015; Noble and Mitchell 2016). Proposed by (Glaser and Strauss 2017), Grounded Theory is commonly used in social and behavioural sciences to uncover social processes: "*...The researcher begins with a broad query in a particular topic area and then collects relevant information about the topic. As the action processes of data collection continue, each piece of information is reviewed, compared, and contrasted with other information. From this constant comparison process, commonalities and dissimilarities among categories of information become clear, and ultimately a theory that explains observations is inductively developed*" (DePoy and Gitlin 2016: 107). Essential features of Grounded Theory include: (1) data collection and analysis occur simultaneously, and (2) categories and analytic codes are developed from data rather than preexisting conceptualisations (Noble and Mitchell 2016).

This research applies Grounded Theory in data collection and analysis in the following ways. First, the researcher spent extensive time in the field gathering data. Second, the data collection is semi-structured and evolving rather than entirely structured. The researcher seeks pluralism and complexity in the data collection process. Third, data is categorised and coded to reduce raw data to themes, a complex and time-consuming process. Fourth, there is extensive writing to show multiple perspectives to find potential patterns and substantiate findings.

4.2.1 Data Collection Methods

Based on Grounded Theory as a methodological philosophy for data collection and analysis, this research uses (1) in-depth semi-structured interviews, (2) archival research, and (3) observations as methods to collect data (see Table 4.2).

Table 4.2 Overview of data collection methods

Data collection methods	Type of data	Sources	Quantity
In-depth interviews	Response from interviewees	Government officials Urban planners Urban governance consultants Academic researchers Land appropriators Communities/ residents	50
Archive research	Urban plan/Policy texts	National environmental agendas National climate plans National urban plans Local environmental agendas Local climate plans Local urban plans	94
Observation	Observation notes	Researcher	/

Table 4.3 Overview of fieldwork

Phases	Dec 2017–Jan 2018 (2 months)	July–Sep 2018 (3 months)	Dec 2018–Feb 2019 (3 months)	End-May–Mid-Jun 2019 and Mid-August 2019 (1 month)
Fieldwork	10 preliminary interviews, archival research, observations	20 formal interviews, archival research, observations	14 formal interviews, archival research, observations	6 follow-up interviews, collect visual data (site images), archival research, and observations

4.2.1.1 Semi-Structured In-Depth Interviews

This research conducts semi-structured in-depth interviews with government officials, urban planners, urban governance consultants, academic researchers, land appropriators, and communities/residents. The aim is to gather focused, qualitative descriptive data about the decision-making processes and outcomes in Guangzhou's three Urban green spaces planning and governance cases. The interviewees are selected based on the principle of maximising the variety of viewpoints: individuals with extensive work experience, insights, or local knowledge from each actor category—government officials, urban planners (including landscape architects), urban governance consultants, academic researchers, land appropriators, and residents are selected and interviewed.

Semi-structured in-depth interviews are an ideal data collection method for this research as it provides the flexibility to follow up on valuable threads that emerged during interviews. Semi-structured interviews are verbal interchanges "*where one person, the interviewer, attempts to elicit information from another person by asking questions ... Semi-structured interviews unfold in a conversational manner offering participants the chance to explore issues they feel are important*" (Longhurst 2010: 103). It allows the researcher to begin with comparatively general questions before inquiring about more specific insights later and allows both parties to clarify queries and terminologies, enriching the research's depth and validity. Besides, during a face-to-face in-depth interview, researchers can observe the non-verbal behaviours of the interviewees, such as body movement and facial expressions, which are informative for the researchers to ask follow-up questions. Furthermore, face-to-face interviews require interviewees to answer questions spontaneously without much time for filtering information or reframing.

Designing semi-structured interview questions

In the previous subchapter, the IAD framework, the variables, and the logic of their interactions are explained. In this subchapter, questions designed for the semi-structured in-depth interviews are structured based on the variables and their interaction in the IAD framework and are primarily open-ended (see Annex 2. Semi-Structured In-Depth Interview Outline) Three major thematic parts are defined to understand the empirical part: (1) land resources and land economics, (2) urban landscape ecology/green space planning, and (3) urban governance and institutions. Based on these three thematic parts and the main research questions, 44 semi-structured interview questions altogether are designed and are oriented for experts in each field:

1. For insights into land resources and land economics: government officials, land economics scholars, and urban governance consultants are interviewed.
2. For insights into urban landscape ecology/green space planning, government officials, urban planning scholars and practitioners, and residents are interviewed.
3. For insights into institutions and governance: government officials and scholars are interviewed.

Questions are designed for the overall urbanisation and institutional contexts of China's urbanisation and the city of Guangzhou, as well as for the specific three case studies chosen in the city of Guangzhou.

Selecting interviewees and preparing for the interviews

The interviewees for this research are chosen based on their knowledge and experience related to the research topic, the socio-economic and political context of China's urbanisation, and the three case studies. 50 people representing different stakeholders/actors in the planning and governance of Urban green spaces in Guangzhou were chosen to participate in the interview. The interview questions are composed of both general and specific questions to capture information in detail and to avoid misunderstanding. The aim of the interviews was to understand how each (group of) stakeholder/actor experience, interact with each other, and make sense of their actions in the underlined circumstance.

1. The initial method of interviewee selection: the researcher began the interviews with academic researchers in the researcher's professional network who were often consulted by the local governments on urban (Urban green spaces) planning issues and were friendly and supportive of this research project.
2. The following method of interviewee selection: the researcher asked the academic researchers who were interviewed to recommend potential interviewees, who could be government officials, urban planning professionals, consultants, or others who are familiar with the context of this empirical research and the three case studies. The request of interviewee recommendations iterated several rounds until the researcher considered that sufficient data was collected for writing this book.
3. Interview preparation: prior to the interviews, the interviewees were invited to sign a form of inform consent for the interview (see Annex 1. Form of Informed Consent for Interview 采访同意书, and to indicate their preference of data recording method (hand notes, audio recording, or video recording). In addition, to enable the most comfortable ambient conditions for the interviewees, most of the interviews took place at the interviewees' work location/office, while a handful of the interviews were conducted via phone calls due to the impossibility of physical in-person meetings.

4.2.1.2 Archival Research

In addition to the qualitative semi-structured interviews, this research also contains an in-depth inquiry into policy and media archives, print and digital, to collect complementary data. Archival evidence can provide the researcher with original and formal policies, plans, and technical reports in details, which are valuable sources to understand policymakers' perceptions of the projects at the time and the cost–benefit calculations within their historical, situational, and communication contexts (Larson 2017). Archival methods are considered beneficial for the robustness, contextuality and holism, particularly of a research endeavour that relates to the study of organisations in multi-disciplinary undertakings (Das et al. 2018).

The archival research conducted for this book contains altogether 88 documents, including strategic plans, conceptual plans, action plans, macro-socio-economic development plans, special plans, government reports, assessment reports, communiqués, and regulations, all of which are effective between 2001 and 2019, the study period of this book (see Annex 5. List of Policy Documents).

1. **strategic plans** entailing policies and actions at national and local levels on climate change, sustainable development (SDGs), public health, ecological protection, urbanisation, and instructional reform
2. **conceptual plans** on how to concretise the concept of *Ecological Civilisation* and the concept of *urban ecological corridor network*
3. **action plans** on air pollution prevention and control at the national level
4. **macro-socio-economic development plans**: including all the Five-Year Plans between 2000 and 2019 at central and municipal levels
5. **special plan** on how to implement *Sponge City* at the local level
6. **government reports** on China's actions and policies on climate change and the work of the local governments of Guangzhou in general
7. **assessment reports** on the implementation of the ecological corridor network in Guangzhou
8. **communiques** of national land greening status during 2001–2019
9. **regulation** on national environmental protection, the latest revision in 2014
10. **technical guidelines** on the implementation of the ecological corridor network and the *Sponge City* program.

These archival data come primarily from four sources:

1. websites of the State Council, ministries, municipal, and district governments
2. physical archives located in government buildings (via in-person visits)
3. physical documents shared/gifted by urban planning professionals exclusively for this study (via in-person meeting exchange).

It is necessary to note that none of the archival data collected is technically confidential; however, some of the documents are made for internal government reference, which has not been released to the public and will not be.

4.2.1.3 Observation

Observation is the systematic description of the events, behaviours, and artefacts of a social setting (Marshall and Rossman 2016: 143). It is a mature method often used in ethnographic, anthropological, and behavioural science studies (Baker 2006; Jersild and Meigs 1939). Observation is an ideal method to understand complex human phenomena, as it permits researchers to study people in their native environment to gain a more comprehensive and holistic understanding of the people and events from the researchers' perspective. It is a context-sensitive method complementary to direct interviews or laboratory measures and can capture aspects related to the social behaviour and preferences of the actors studied (Baker 2006; DeWalt and DeWalt 2010; Prus and Jorgensen 1990). Using observation, the researcher can obtain firsthand insights into the mundane detail of interviewees' behaviours, attitudes, and language. It allows the researcher to learn about issues the interviewee may be unaware of, are unwilling, or unable to discuss candidly during an interview (Frechtling et al. 2002).

This book uses observation as a complementary method to the two other qualitative data collection methods—semi-structured in-depth interviews and archival research. To the best ability, the observation is conducted respecting the principles of authenticity, trustworthiness, reflexivity, and particularity (Patton 2015). The observation is conducted unstructured in natural settings alongside semi-structured interviews, in which the researcher and the interviewees have established informed consent and trust. The observation data obtained for this research exists primarily in the form of field notes, which contain descriptions of the interview activities, interactions, and behaviours of interviewees, actions of each actor, and processes of the three cases from planning to implementation and management. The field notes are then coded and analysed together with the interview transcripts.

4.2.2 *Ethical Considerations*

- Data confidentiality: interviewees were invited to participate in the semi-structured, in-depth interviews voluntarily and were ensured that their answers would be kept confidential and used solely for the purpose of this research. All the interview data was collected, either in audio recording or just handwritten notes depending on the interviewees' preference, and then stored and treated in a confidential, anonymous, and codified way.
- Protection and anonymisation of interviewees: the interviewees were invited to sign a consent form (see Annex 1. Form of Informed Consent for Interview 采访同意书) prior to the interviews with the researcher.
- Data quality control: as introduced in the data collection methods above, there are multiple data sources and multiple methods to strengthen the credibility and robustness of the data. Besides, all the interviewees are invited to sign a form of consent prior to participating in the interviews, indicating their preferences of data

collection (via audio recording or only handwritten notes). Prior to the interviews, they are also guaranteed that the interview subjects will be non-sensitive, non-discriminative, and non-incriminating. Furthermore, this book contains a substantial chapter of contextualisation of the empirical study (see Chapter V), in order to explain why these kinds of qualitative data are collected (Krippendorff 2018).

4.2.3 *Data Collection Timeframe*

The researcher has spent an extensive amount of time in the field study location, Guangzhou, China, to gather data to ensure the depth, plurality, and complexity of the data for this qualitative case-based research. This includes four phases of data collection of nine months in total, between December 2017–August 2019.

4.3 Data Analysis

4.3.1 *Content Analysis Using MaxQDA*

4.3.1.1 Qualitative Content Analysis

Content analysis is "a research technique for making replicable and valid inferences from texts (or other meaningful matter) to the contexts of their use" (Krippendorff 2004: 18). Content analysis is a scientific tool that assists the researcher to understand phenomena starting by recognising meanings and then systematically analysing them within a particular set of texts. The IAD framework presented above as the analytical framework for this research and the logic of the research questions are present throughout the process of content analysis. Qualitative content analysis in the eye of a reader often has the characteristic of extensive description and discussion. This is necessary to show multiple perspectives as well as to substantiate findings, not only within the qualitative data material gathered but also the context in which the data was gathered (Creswell 1998; Krippendorff 2004).

The aim of the qualitative content analysis in this book is to (1) understand the institutional dynamics, facilitators, and underlying barriers Chinese local governments have faced regarding the need and mandates to plan and manage Urban green spaces that could lead to more effective Nature-based Solutions, (2) examine to which extent local governments have been able to advance and upscale Urban green spaces and with what approaches, (3) analyse the costs and benefits that Chinese local governments have taken into consideration for planning and managing Urban green spaces, and (4) identify intrinsic-value-related factors that have limited Chinese local governments to pursue greater Urban green spaces quantity and quality, and effectiveness of Nature-based Solutions.

4.3.1.2 Using MaxQDA Qualitative Data Analysis Software

This research uses MaxQDA, a Qualitative Data Analysis Software (QDA software), for organising and analysing the qualitative data collected from semi-structured interviews, policy archival research, and observation field notes. MaxQDA has part of its name "Max" associated to the sociologist Max Weber and is rooted in social science methodology. The software provides functions for transcribing, analysing, and visualising qualitative data; hence, it can analyse all the data commonly collected in the context of empirical social science research (Kuckartz and Rädiker 2019). A central feature of MaxQDA is the option of working with codes (categories) and assigning codes to selected parts of the data in order to organise large volumes of text. As the empirical data in this social science research is qualitative and in large volumes, MaxQDA is therefore a suitable tool for data analysis.

4.3.1.3 Organising Data and Coding

The data collected in this research includes information gathered through semi-structured interviews and from archival research, and observations field notes. The researcher used the mobile application "Voice Memos" as a tool to record the interviews when applicable. Some interviewees only agreed to the form of handwritten notes for recording the interview contents. Below are the steps of using MaxQDA software to process the data collected.

1. The first step is to transcribe and organise the audio recordings and handwritten notes. In addition to full transcription files, a summary for each transcription is also produced to organise and navigate effectively inside the software.
2. The second step is to categorise data and define variables according to the IAD framework. In this data analysis, there are seven categories, and they are the same as the first seven variables of the IAD framework: ① *biophysical conditions*, ② *community attributes*, ③ *rules-in-use*, ④ *action situation*, ⑤ *actors*, ⑥ *interactions*, and ⑦ *outcome*.
3. The third step is to code the data within each category. The codes are generated by sorting and sifting through the text within each category. Codes are created when thematic commonalities are identified. Most of the codes have a set of subcodes, or in other words, the coded themes have a set of subthemes which are even more specific. For example, the first category—① biophysical conditions—contains three codes or coded themes, which are then further distinguished with a set of subcodes or subcoded themes (see Table 4.4: Categories, codes, and subcodes for the qualitative content analysis using MaxQDA).

These codes and subcodes serve to identify and mark the relevant texts for the content analysis. Furthermore, the codes for the last two categories—⑥ *interactions* and ⑦ *outcome*—are matched with the respective criteria in the multi-criteria assessment (MCA) for the purpose of conducting the MCA. This will be explained in the following Sect. 4.3.2 Multi-Criteria Assessment (MCA). For a complete view on the

Table 4.4 Categories, codes, and subcodes for the qualitative content analysis using MaxQDA

Categories	Codes (themes)	Subcodes (subthemes)
① Biophysical conditions	(1) Natural condition of Guangzhou	(1) Natural landscape (2) Location (3) Climate
	(2) Climate change impacts on Guangzhou	(4) Types of impacts (5) Population impacted (6) Future trends
② Community attributes	(3) Socio-economic conditions of Guangzhou	(7) Status of urbanisation (8) Status of population growth (9) Status of urban land-use
	(4) Environmental characteristics of the areas	(10) Environmental characteristics of the areas
	(5) Socio-economic characteristics of communities	(11) Socio-economic characteristics of communities
	(6) Urban planning status of the areas	(12) Urban planning status of the areas
③ Rules-in-Use	(7) Governance structure	(13) Interdepartmental positionings (14) Central-local dynamics (15) Public–private relationships (16) Public-civil responsibilities
	(8) Traditional political philosophy	(17) Traditional political philosophy
	(9) Contemporary political philosophy	(18) Socialism with Chinese Characteristics (19) Scientific Development (20) Ecological Civilisation
	(10) Environmental protection history	(21) Pre-2000 (22) 2000 – pre-Paris Agreement (23) Paris Agreement - present
	(11) Land-use and spatial planning system	(24) Typology of urban plans (25) Institutions of urban planning (26) Process of urban planning
	(12) Property rights arrangements	(27) Right to access (28) Right to withdraw (29) Right to manage (30) Right to exclude (31) Right to alienate
④ Action situation	(13) Action situation	(32) Action situation in Case 1 (33) Action situation in Case 2 (34) Action situation in Case 3

(continued)

Table 4.4 (continued)

Categories	Codes (themes)	Subcodes (subthemes)
⑤ Actors	(14) Types of actors	(35) Guangzhou Municipal Government (36) Panyu District Government (37) Haizhu District Government (38) Tianhe District Government (39) Farmland owners (40) Urban migrants (41) Small (informal) businesses (42) Ordinary citizens
	(15) Information known by actors	(43) Known environmental costs (44) Known environmental benefits (45) Known social costs (46) Known social benefits (47) Known economic/financial costs (48) Known economic/financial benefits (49) Other known costs (50) Other known benefits
	(16) Information unknown by actors	(51) Information unknown by actors
⑥ Interactions	(17) Informing/informed	(52) Not informed and not participating (53) Informed but not participating
	(18) Partially participating	(54) Agreeing (55) Coordinating (56) Cooperating (57) Negotiating (58) Supervising
	(19) Substantially participating	
⑦ Outcome	(20) Plans/policies/Urban green spaces implementation	(59) Urban green spaces goals from plans/policies (60) Implemented Urban green spaces
	(21) Changes in institutions	(61) Informal changes (norms, values) (62) Formal changes (new rules, laws, regulations)
	(22) Implications	(63) Climate resilience (64) Human health and well-being (65) Urban climate governance

seven categories, 22 codes (themes) and 65 subcodes (subthemes), please refer to Table 4.4: Categories, codes, and subcodes for the qualitative content analysis using MaxQDA.

4.3.2 *Multi-criteria Assessment (MCA)*

4.3.2.1 Multi-criteria Assessment (MCA) for Qualitative Research

MCA is a "*structured approach used to determine overall preferences among alternative options, where the options accomplish several objectives*" endorsed by the UNFCCC for climate adaptation planning and practices (Bell et al. 2003; UNFCCC 2002). MCA can help researchers understand dynamics between stakeholders during interactions and cost–benefit considerations or other policy trade-offs, as well as the quality of the outcome. In MCA, attributes and indicators are first identified and then measured by monetary and non-monetary dimensions. In some cases, MCA can also be achieved by measuring qualitative characteristics by quantifying their significance (e.g., weighting, scoring, and ranking). MCA is frequently used for estimating the quality of governance by assessing the decision-making processes and the social, economic, and environmental impacts of relevant policies.

A similar method conventionally used by scholars is the cost–benefit analysis method, a comparatively more straightforward method used for project evaluation and can be viewed as a part of the MCA method. Scholars have argued that cost–benefit analysis could enhance understanding of urban greening benefits as people are more amenable to monetary than ecological values; the benefits and amenities could be expressed in dollars (Jim 2013; Nowak and Dwyer 2011). As a matter of fact, a few indirect cost–benefit analysis methods have been applied to assess the economics of Urban green spaces as they are not a marketable commodity. For example, the contingent valuation method explores citizens' willingness-to-pay to use Urban green spaces to prevent Urban green spaces loss or preserve urban biodiversity (Chen and Jim 2010; Lo and Jim 2012). Another method is called the hedonic pricing method, which has been applied to analyse and estimate property transaction prices attributed to urban greening and other natural areas (Garrod and Willis 1994; Jim and Chen 2007). The problem with these methods is that they cannot adequately address the complexity and transboundary characteristic of modern urban mobility (particularly for a rapidly growing megacity like Guangzhou) and the potential of the ecosystem services of Urban green spaces for climate mitigation and adaptation at the local level.

Compared to cost–benefit analysis, the MCA method can capture dimensions outside the range of direct economic appropriability effectively (Samuelson and Nordhaus 2010). For example, land as a resource of scarcity, its fertility may be appropriately captured by the farmer who sells rice or apples produced on the land, in which case the land resource is called an appropriable resource. However, the vegetation and the ecosystem which are supported by land, the benefits brought by their ecosystem services such as provision, regulating, habitat, or supporting so far cannot be accurately calculated in monetary terms. In this circumstance, the MCA method can address the indirect costs associated with Urban green spaces, e.g., the lack of Urban green spaces could lead to chronic health and well-being issues (highlighted during the COVID-19 pandemic, see Liu 2021).

The MCA method is increasingly used in climate adaptation planning to evaluate better the risk and uncertainty dimensions of climate impacts on the social, environmental, technical, and economic spheres. MCA is, therefore, more comprehensive than conventional cost–benefit analysis and can provide more extensive information to support policy-making (Chambwera et al. 2015). For example, *equity* is a crucial dimension to consider in climate adaptation planning, which comprises the gains and losses and impacts on individual citizens and communities, especially vulnerable groups. Many of these gains and losses and impacts have low monetary calculability. MCA has been applied to an increasing number of climate change adaptation studies, including urban flood risk (Chambwera et al. 2015).

4.3.2.2 Applying Multi-criteria Assessment in This Research

Developing criteria

The criteria used in this assessment (see Table 4.5) are developed taking references from (1) the ⑧evaluative criteria in the IAD framework, (2) the universal principles for effective governance in the *Millennium Ecosystem Assessment (MEA)*, (3) the *Technical Assessment Synopsis Reports* of the European Green Capital Award,[1] and (4) the multi-criteria evaluation (MCE) guideline[2] for ENavi—the Energy Transition Navigation System developed under the German Energiewende national energy transition scheme.

First, taking the ⑧ evaluative criteria in the IAD framework as the initial reference helps to maintain the logical structure of the analysis as well as the theoretical rigour of this research. These evaluative criteria are: (1) *Economic efficiency*, (2) *equity*, (3) *adaptability, resilience, and robustness*, (4) *accountability*, (5) *and conformance to general morality* (for more detailed definitions of these five criteria, please revisit Sect. 3.5 Analysing Common-Pool Resource Governance Through the IAD Framework). Despite that the IAD framework has been well acclaimed by scholars for its overall potential in analysing CPR governance, the ⑧ evaluative criteria are the least developed part subjected to future contestation as scholar research advances (Cole 2017). Considering that this is also the most flexible part of the IAD framework that allows context-specific readjustments, it is therefore used as an initial reference.

Second, the *Millennium Ecosystem Assessment*´s four universal core principles of effective governance contribute to the MCA particularly on the dynamics between governments and non-state actors during interactions, thereby can complement the

[1] https://ec.europa.eu/environment/europeangreencapital/wp-content/uploads/2013/02/EGCA_2020_Technical_Assessment_Synopsis_Report.pdf.

[2] Quitzow (2019).

Table 4.5 MCA criteria for this research—An overview of references

Initial and structural reference	Complementary references			
IAD framework	Millennium ecosystem assessment	European green capital award	Guangzhou award for urban innovation	ENavi—Energiewende
1. Economic efficiency 2. Equity 3. Adaptability, resilience, and robustness 4. Accountability 5. Conformance to general morality	1. Compliance 2. Legitimacy 3. Interlinkages between institutions 4. Adequate financing	1. Climate change mitigation 2. Climate change adaptation 3. Sustainable urban mobility 4. Sustainable land-use 5. Nature and Biodiversity 6. Air quality 7. Noise 8. Waste 9. Water 10. Green growth and eco-innovation 11. Energy Performance 12. Governance	1. Innovativeness 2. Effectiveness 3. Replicability and transferability 4. Significance and relevance to the SDGs and the NUA	**Evaluating targets** 1. Effectiveness 2. Cost efficiency 3. Total costs 4. Resilience (as a more broadly defined form of supply security) **Evaluating "side effects"** 5. Economic predictability and contribution to the common good 6. Health and environmental protection and resource conservation 7. Fostering social cohesion **Evaluating compatibility with legal, political, and ethical norms and values** 8. Legitimacy 9. Ethical acceptability 10. Legality

initial reference. These four principles are: (1) compliance, (2) legitimacy, (3) interlinkages between institutions, and (4) adequate financing. Compliance describes the degree to which governments follow formal rules and obligations imposed by a higher-level government (also considered a transaction cost factor). Legitimacy includes the clarity of the rules, the interpreted coherence of the rules and their adherence to the existing hierarchy of rules (e.g., if the current Urban green spaces planning in Guangzhou adheres to national environmental agendas such as *Ecological Civilisation*). And then, the interlinkages between institutions also influence the effectiveness of governance, as often several institutions work together to formulate and implement urban plans to ensure effectiveness. Furthermore, adequate financing is important for supporting proper implementation (Fisher et al. 2005).

Third, the European Green Capital Award (EGCA) is an initiative launched by the European Commission aiming to recognise cities leading the way on environmentally friendly urban living. Hence, it is also an adequate reference to complement the MCA, particularly from an urban planning practice and governance perspective.

Fourth, the Guangzhou International Award for Urban Innovation (the Guangzhou Award)[3] is established to recognise innovation in projects, measures or policies in cities and local governments that improve social, economic, and environmental sustainability. The Award is hosted by the City of Guangzhou and co-sponsored by the United Cities and Local Governments (UCLG), the World Association of the Major Metropolises (Metropolis). The criteria applied to the assessment for the Award include (1) innovativeness, (2) effectiveness, (3) replicability and transferability, (4) significance and relevance to the SDGs and the New Urban Agenda (NUA). This reference adds a more context-specific reference on urban governance (Bina 2008a, b); hence, it is also supportive for developing the criteria for the MCA in this research.

Lastly, the MCE guideline for analysing ENavi (the Energy Transition Navigation System developed under the German Energiewende national energy transition scheme) was developed by the Potsdam Institute of Advanced Sustainability Studies (IASS) in 2018. This evaluation guideline includes ten criteria divided into three groups highlighting the many different aspects of sustainability. Therefore, this reference is also competent for complementing the MCA in this research.

Building based on the structural reference from the IAD framework and the complementary references, a new set of criteria is developed for the MCA in this research (see Table 4.6: Evaluative criteria developed for the multi-criteria assessment in this research). The new set of criteria absorbed all the relevant dimensions and indicators from the references above and is defined as the second-tier criteria. There are altogether 28 criteria covering the (1) economic, (2) technical, (3) ecological, (4) social, and (5) institutional dimensions for the empirical case study in this research.

Evaluating interactions and outcomes

As mentioned in Sect. 4.3.1.3 Organising data and coding, the codes (subthemes) for the categories—⑥ interactions and ⑦ outcome—are matched with the respective criteria in MCA. The match is presented in Table 4.7: MCA criteria matched with the ⑥ interactions and ⑦ outcomes of three case studies for evaluation.

A 1–7 scale evaluation scheme is used with respective colours assigned for filling Table 4.7: MCA criteria matched with the ⑥ interactions and ⑦ outcomes of three case studies for evaluation: 1—Very clearly negative, 2—Clearly negative, 3—Slightly negative, N/A—Not sufficient information to assess, 4—Neutral/Insignificant, 5—Slightly positive, 6—Clearly positive, 7—Very clearly positive.

[3] The Guangzhou International Award for Urban Innovation. Available at: http://www.guangzhouaward.org/.

Table 4.6 Evaluative criteria developed for the multi-criteria assessment in this research

First-tier and structural criteria	Dimensions	Second-tier criteria
1. Economic efficiency	Economic	1. Financing adequacy 2. Direct cost–benefit efficiency
	Technical	3. Temporal efficiency of planning 4. Temporal efficiency of implementation
	Institutional	5. Interlinkages between institutions
2. Equity	Social	6. Cultivation of social cohesion 7. Inclusion of health and well-being benefits 8. Inclusion of rights to access information 9. Inclusion of rights to monitor 10. Inclusion of rights to participate
	Ecological	11. Environmental protection and resource conservation
	Economic	12. Inclusion of economic benefits
3. Adaptability, resilience, and robustness	Technical/ecological	13. Sustainability in land-use 14. Environmental quality (air, noise, waste, water) 15. Ecological resilience (ability to stay resilient) 16. Effectiveness in climate change mitigation and adaptation 17. Significance and relevance to the SDGs 18. Technical replicability
	Economic	19. Economic predictability and contribution to the common good 20. Green growth and eco-innovation
	Institutional	21. Innovativeness in governance 22. Institutional adaptability (changes over time) 23. Transferability of practices
4. Accountability	Institutional	24. Compliance with upper-level governments and agendas
		25. Administrative accountability within own level
5. Conformance to general morality	Institutional	26. Information transparency
		27. Legitimacy (conformity to the law)
		28. Ethical acceptability

Table 4.7 MCA criteria matched with the ⑥ interactions and ⑦ outcomes of three case studies for evaluation

First‾tier and structural criteria	Dimensions	Second-tier criteria	First case	Second case	Third case	Corresponding subcodes (subthemes)	Corresponding categories
1. Economic efficiency	Economic	1. Financing adequacy 2. Direct cost–benefit efficiency				Urban green spaces goals from plans/policies Urban green spaces implementation	⑦ **Outcome**
	Technical	3. Temporal efficiency of planning 4. Temporal efficiency of implementation				Urban green spaces goals from plans/policies Urban green spaces implementation	⑦ **Outcome**
	Institutional	5. Interlinkages between institutions				Urban green spaces goals from plans/policies Urban green spaces implementation Urban climate governance	⑦ **Outcome**
2. Equity	Social	6. Cultivation of social cohesion				Urban green spaces implementation Informing/informed Partially participating Substantially participating	⑦ **Outcome** ⑥ **Interactions**
		7. Inclusion of health and well-being benefits				Human health and well-being	⑦ **Outcome**
		8. Inclusion of rights to access information 9. Inclusion of rights to monitor 10. Inclusion of rights to participate				Urban green spaces implementation Informing/informed Partially participating Substantially participating	⑦ **Outcome** ⑥ **Interactions**
	Ecological	11. Environmental protection and resource conservation				Urban green spaces implementation Climate resilience	⑦ **Outcome**

(continued)

Table 4.7 (continued)

First‾tier and structural criteria	Dimensions	Second-tier criteria	First case	Second case	Third case	Corresponding subcodes (subthemes)	Corresponding categories
	Economic	12. Inclusion of economic benefits				Urban green spaces goals from plans/policies Urban green spaces implementation	⑦ **Outcome**
3. Adaptability, resilience, and robustness	Technical/ ecological	13. Sustainability in land-use				Urban green spaces implementation Climate resilience Human health and well-being	⑦ **Outcome**
		14. Environmental quality (air, noise, waste, water)					
		15. Ecological resilience (ability to stay resilient)					
		16. Effectiveness in climate change mitigation and adaptation					
		17. Significance and relevance to the SDGs					
		18. Technical replicability					
	Economic	19. Economic predictability and contribution to the common good				Urban green spaces goals from plans/policies	⑦ **Outcome**
		20. Green growth and eco-innovation					
	Institutional	21. Innovativeness in governance				Urban green spaces goals from plans/policies Urban green spaces implementation Formal changes Informal changes Urban climate governance	⑦ **Outcome**
		22. Institutional adaptability (changes over time)					
		23. Transferability of practices					

(continued)

Table 4.7 (continued)

First¯tier and structural criteria	Dimensions	Second-tier criteria	First case	Second case	Third case	Corresponding subcodes (subthemes)	Corresponding categories
4. Accountability	Institutional	24. Compliance with upper-level governments and agendas				Urban green spaces goals from plans/policies Urban green spaces implementation Urban climate governance	⑦**Outcome**
		25. Administrative accountability within own level				Urban green spaces goals from plans/policies Urban green spaces implementation Urban climate governance	⑦ **Outcome**
5. Conformance to general morality	Institutional	26. Information transparency				Urban green spaces goals from plans/policies Urban green spaces implementation	⑦ **Outcome**
		27. Legitimacy (conformity to the law)				Urban green spaces goals from plans/policies Urban green spaces implementation Formal changes Informal changes	⑦ **Outcome**
		28. Ethical acceptability				Urban green spaces goals from plans/policies Urban green spaces implementation Formal changes Informal changes	⑦ **Outcome**

Table 4.8 1–7 scale evaluation scheme for the MCA

1	Very clearly negative	7	Very clearly positive
2	Clearly negative	6	Clearly positive
3	Slightly negative	5	Slightly positive
N/A	Not sufficient information to assess	4	Neutral/Insignificant

References

Baker LM (2006) Observation: a complex research method. Libr Trends. https://doi.org/10.1353/lib.2006.0045

Bell ML, Hobbs BF, Ellis H (2003) The use of multi-criteria decision-making methods in the integrated assessment of climate change: implications for IA practitioners. Socioecon Plann Sci. https://doi.org/10.1016/S0038-0121(02)00047-2

Bina O (2008a) Context and systems: thinking more broadly about effectiveness in Strategic Environmental Assessment in China. Environ Manage. https://doi.org/10.1007/s00267-008-9123-5

Bina OC (2008) Relevance of Europe's Environmental Policy Integration concept for China: evidence of de facto EPI and questions about effectiveness (No. 37; EPIGOV/CITIZENS-2004-4.2.2—Governance for Sustainable Development).

Chambwera M, Heal G, Dubeux C, Hallegatte S, Leclerc L, Markandya A, McCarl BA, Mechler R, Neumann JE, Calvo E, Iglesias A, Navrud S, Kairiza T (2015) Economics of adaptation. In: Climate change 2014 impacts, adaptation and vulnerability: Part A: global and sectoral aspects. https://doi.org/10.1017/CBO9781107415379.022

Charmaz K (2015) Grounded theory: methodology and theory construction. In: International encyclopedia of the social & behavioral sciences, 2nd edn. https://doi.org/10.1016/B978-0-08-097086-8.44029-8

Chen WY, Jim CY (2010) Resident motivations and willingness-to-pay for urban biodiversity conservation in Guangzhou (China). Environ Manage. https://doi.org/10.1007/s00267-010-9478-2

Cole DH (2017) Laws, norms, and the Institutional Analysis and Development framework. J Inst Econ. https://doi.org/10.1017/S1744137417000030

Creswell JW (1998) Qualitative inquiry and research design: choosing among five traditions. Qual Health Res

Cumming GS, Cumming DHM, Redman CL (2006) Scale mismatches in social-ecological systems: causes, consequences, and solutions. Ecol Soc. https://doi.org/10.5751/ES-01569-110114

Das R, Jain KK, Mishra SK (2018) Archival research: a neglected method in organization studies. Benchmarking. https://doi.org/10.1108/BIJ-08-2016-0123

Denzin NK, Lincoln YS (2000) Introduction: the discipline and practice of qualitative research. In: Handbook of qualitative research, 2nd edn.

DePoy E, Gitlin LN (2016) Formulating research questions and queries. Introduct Res. https://doi.org/10.1016/b978-0-323-26171-5.00008-2

DeWalt KM, DeWalt BR (2010) Informal interviewing in participant observation. In: Participant observation: a guide for fieldworkers.

Fisher DR, Simpson RD, Vira B, Chambers WB, Davidson DJ (2005) Millennium ecosystem assessment volume 3: ecosystems and human well-being: policy responses—Chapter 3: assessing responses. https://millenniumassessment.org/en/Responses.html#download

Frechtling J, Frierson Hood S, Hughes G, Katzenmeyer C (2002) The 2002 user-friendly handbook for project evaluation. The National Science Foundation, Directorate for Education and Human Resources, Division of Research, Evaluation, and Communication. https://www.nsf.gov/pubs/2002/nsf02057/nsf02057.pdf

Garrod G, Willis K (1994) An economic estimate of the effect of a waterside location on property values. Environ Resour Econ. https://doi.org/10.1007/BF00692204

Glaser BG, Strauss AL (2017) Discovery of grounded theory: strategies for qualitative research. Routledge. https://books.google.pt/books?id=GTMrDwAAQBAJ&dq=The+Discovery+Of+Grounded+Theory&lr=&source=gbs_navlinks_s

Guangzhou Daily (2018) Construction of ecological corridors in guangzhou: four districts completed (In Chinese: 广州生态廊道建设 四区实施方案完成). http://m.xinhuanet.com/gd/2018-04/23/c_1122724389.htm

Jersild AT, Meigs MF (1939) Chapter V: direct observation as a research method. Rev Educ Res 9(5):472–482. https://doi.org/10.3102/00346543009005472

Jim CY (2013) Sustainable urban greening strategies for compact cities in developing and developed economies. Urban Ecosyst 16(4):741–761. https://doi.org/10.1007/s11252-012-0268-x

Jim CY, Chen WY (2007) Consumption preferences and environmental externalities: a hedonic analysis of the housing market in Guangzhou. Geoforum. https://doi.org/10.1016/j.geoforum.2006.10.002

Krippendorff KH (2004) *Contenet analysis—An introduction to its methodology*, 2nd ed. Sage Publications, Inc.

Krippendorff K (2018) Content analysis—An introduction to its methodology, 4th ed. SAGE Publications Inc. https://us.sagepub.com/en-us/nam/content-analysis/book258450

Kuckartz U, Rädiker S (2019) Analyzing qualitative data with MAXQDA. In: Analyzing qualitative data with MAXQDA. https://doi.org/10.1007/978-3-030-15671-8

Larson DW (2017) Archival research in foreign policy. In: Oxford research encyclopedia of politics. Oxford University Press. https://doi.org/10.1093/acrefore/9780190228637.013.389

Liu J (2021) COVID-19 shows us the need to plan urban green spaces more systemically for urban health and wellbeing. In Gatzweiler FW (ed) Urban health and wellbeing programme policy briefs, vol 2. Zhejiang University Press and Springer, pp 39–44

Lo AYH, Jim CY (2012) Citizen attitude and expectation towards greenspace provision in compact urban milieu. Land Use Policy. https://doi.org/10.1016/j.landusepol.2011.09.011

Longhurst R (2010) Semi-structured interviews and focus groups. J Chem Inf Model

Marshall C, Rossman GB (2016) In: Salmon H (ed) Designing qualitative research, 6th edn. SAGE Publications, Inc. https://books.google.pt/books?hl=en&lr=&id=-zncBQAAQBAJ&oi=fnd&pg=PP1&dq=Marshall,+C.+%26+Rossman,+G.+B.+(1989).+Designing+qualitative+research.+Newbury+Park,+CA:+Sage.&ots=Lgf6oEVwcY&sig=vThLBgIMTsOV_YgMqRgfFqnwlHo&redir_esc=y#v=onepage&q&f=false

Noble H, Mitchell G (2016) What is grounded theory? Evid Based Nurs 19(2):34–35. https://doi.org/10.1136/eb-2016-102306

Nowak DJ, Dwyer JF (2011) Understanding the benefits and costs of urban forest ecosystems. In: Handbook of urban and community forestry in the northeast. https://doi.org/10.1007/978-1-4615-4191-2_2

Patton MQ (2015) Qualitative research & evaluation methods—Integrating theory and practice, 4th edn. SAGE Publications Inc. https://us.sagepub.com/en-us/nam/qualitative-research-evaluation-methods/book232962

Prus R, Jorgensen DL (1990) Participant observation: a methodology for human studies. Can J Sociol/cahiers Canadiens De Sociologie. https://doi.org/10.2307/3340930

Quitzow (2019) A sustainable energy transition: new guidelines for multi-criteria evaluation. https://www.iass-potsdam.de/en/news/energiewende-guidelines-multi-criteria-evaluation

Samuelson PA, Nordhaus WD (2010) Microeconomics, 19th edn. McGraw-Hill/Irwin

Schlüter A (2010) Institutional change and qualitative research methodological considerations for institutional economic empirical research. J Interdiscip Econ. https://doi.org/10.1177/02601079X10002200405

State Council of China (2001) Outline of the People's reputenth five-year plan for national economic and social development of the blic of China (In Chinese: 中华人民共和国国民经济和社会发展第十个五年计划纲要). Available at: https://www.gov.cn/gongbao/content/2001/content_60699.htm

UNFCCC (2002) Annotated guidelines for the preparation of national adaptation programs of action, p 41. UNFCCC Secretariat. https://unfccc.int/files/cooperation_and_support/ldc/application/pdf/annguide.pdf

Young O (2002) The institutional dimensions of environmental change: fit, interplay, and scale. Organization. https://doi.org/10.1023/A:1024850005580

Chapter 5
Empirical Context and Case Presentation

5.1 Biophysical Contexts of China and of Guangzhou City

5.1.1 Natural and Socio-economic Conditions

China has the third-largest land area globally with about 9.6 million km^2, stretching 5000 km from east to west and 5500 km from north to south with widely varying landscapes, including mountains, high plateaus, sandy deserts, dense forests, coastal wetlands, plain fertile agricultural fields, and more. The country is located on the west coast of the Pacific Ocean with a continental coastline of more than 18,000 km and a complex and diverse range of climates characterised by monsoon with high temperatures and rain in summers, cold and little rain in winters. The high-temperature periods are often consistent with the rainy period.

In addition, China is currently the most populous country with 1.39 billion population and four of the ten largest metropolitan cities globally, each hosting well over 20 million people. Meanwhile, the country has the second-largest economy distributed primarily along the eastern coastline, particularly in the Yellow River Delta (YRD) and the Pearl River Delta (PRD). The drastic divide between the eastern and western sides of the country, both in biophysical and socio-economic conditions, is often marked by the reference to the Heihe-Tengchong Line (see Fig. 5.1: The Heihe-Tengchong Line an imaginary northeast-southwest line proposed by Chinese geographer Hu Huanyong as a geo-demographic demarcation line. This line reveals that the eastern side of China contains about 36% of China's total land and is semi-humid but supports around 96% of the total population, while the western side of China marked this line contains up to 64% of the total land and is semi-arid, but only supports 4% of the population (Pan 2015). Considering such a high population density on the eastern side of the country, the socio-economic needs are pushing the ecological capacity of the eastern side to the limit. According to the *Report on China's Ecological Footprint* co-released by the World Wildlife Fund (WWF) and

© Zhejiang University Press 2023

J. Liu, *Transforming Urban Green Space Governance in China Under Ecological Civilization: An Institutional Analysis*, Urban Health and Wellbeing,
https://doi.org/10.1007/978-981-99-6694-3_5

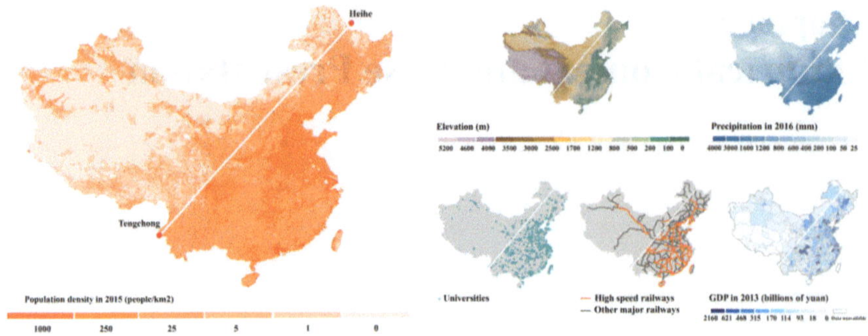

Fig. 5.1 Heihe-Tengchong line. *Source* Murphy (2017)

the China Council for International Cooperation on Environment and Development (CCICED) in 2012, provinces in China that have ecological deficits (when the footprint of a population exceeds the biocapacity of the area available to that population) grew from 19 to 26 between 1980 and 2000. The speed of change drastically outpaced the world average, primarily due to the rapid increase in both intensive agricultural production and fossil fuel consumption (Cheng et al. 2012).

Guangzhou is the third-largest metropolis in China and the economic, political, and cultural centre of Guangdong province located in the southeast close to the low-lying composite Pearl River Delta 0.3–0.4 m below mean sea level (MSL) (Tracy et al. 2007) (see Figs. 5.2 and 5.4). Guangzhou has a total urban territory of 7434 km^2 between 112.57–114.3° E and 22.26–23.56° N and a subtropical oceanic monsoon climate[1] characterised by hot summers and warm winters. The city receives abundant rainfall of about 150 days a year and average annual precipitation of more than 1800 mm, most of which during the flood season beginning in April and peaks in June. Tropical cyclones tend to visit in summers (June–August), bringing abundant precipitation. The hottest month of the year usually is July.[2]

Rapid urbanisation and land-use change

Since the national economic reform in the early 1980s, the total built area in Guangzhou at least tripled by 2010 (Zhou 2010) (for a transformative example, see Fig. 5.5). Before 2000, Guangzhou's urban area only contained five districts: Yuexiu, **Haizhu**, Liwan, **Tianhe**, and Baiyun (see Fig. 5.3, marked in red letters). The other six districts: Huangpu, Huadu, Panyu, Nansha, Conghua, and Zengcheng, were all established or merged into Guangzhou after 2000. **Panyu** and Huadu were merged into Guangzhou as individual districts in 2000. The District of Nansha was established in 2005. Conghua and Zengcheng were merged into Guangzhou as individual districts in 2014, and Luogang District was merged into Huangpu District to

[1] Chinese National Bureau of Technical Supervision (1994).

[2] Homepage of the Municipal Government of Guangzhou > city information > overview of Guangzhou > Natural Geography. Available at: http://www.gz.gov.cn/gzgov/s2769/sq_tt.shtml.

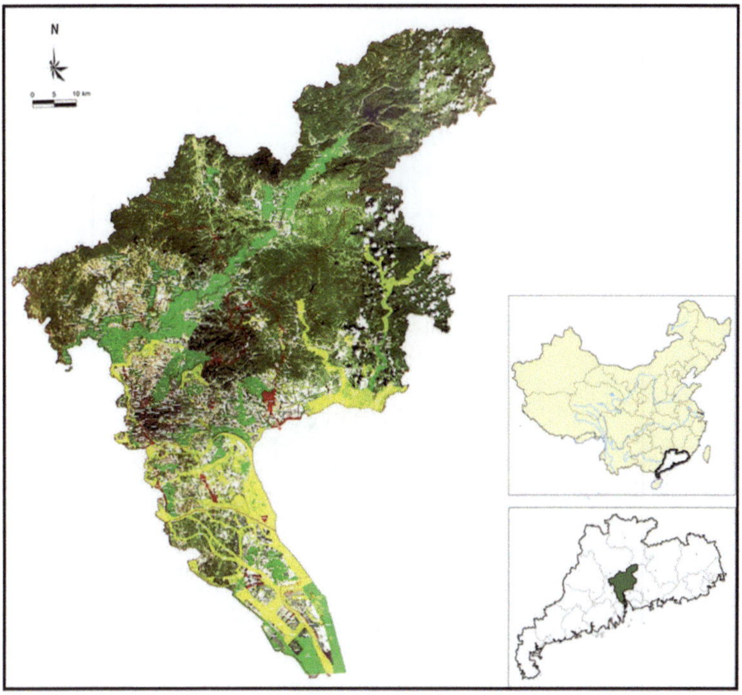

Fig. 5.2 Geographic location of Guangzhou

become the new Huangpu District in the same year. Today, Guangzhou's jurisdictional boundary embraces 11 districts, including Yuexiu, **Haizhu**, Liwan, **Tianhe**, Baiyun, Huangpu, Huadu, **Panyu**, Nansha, Conghua, and Zengcheng occupying a total land area of 7434.4 km^2 (see Fig. 5.3).

The rapid and extensive industrialisation and urbanisation Pearl River Delta area experienced following the economic reform in the early 1980s plus the historically lagged land property rights confirmation work resulted in the overall low land-use efficiency and evident urban land scarcity by the New Millennium. For example, by 1989, five years into the implementation of the 14th Urban Master Plan of Guangzhou (1984–2000), the city had almost used all of the urban land quotas for the planned period (Dai and Duan 2003). The appeal for urban land expansion was extremely evident during the subsequent planning work for the 15th edition of the Urban Master Plan of Guangzhou (Deng et al. 2017). The increasing urban land scarcity and land needs in the vibrant economic development prompted the city to expand by merging neighbour districts or cities. The landscape of Guangzhou experienced great changes between 1990 and 2014 with the tendency likely to continue (Gong et al. 2018). In 2018, the central government proposed to integrate Hong Kong, Macau and nine southern mainland cities into one huge megacity, the largest megacity hub that the

Fig. 5.3 Current administrative districts and their distribution of Guangzhou. *Note* The red texts indicate the administrative districts established in or before 2000, and the black letters indicate the districts established or merged after 2000

world has seen, alias the Guangdong-Hong Kong-Macau Greater Bay Area (GD-HK-MO GBA), with the vision of turning a leading hub for innovation and economic growth.

Population and urbanisation in Guangzhou

Guangzhou's residential population grew from 5.9 million in 1990 to 14 million in 2016, and the urban population accounts for 86%. Urban population is the number of residents who are registered as urban households. The urban and rural household registrations are a unique system in China which are invented initially to mediate population movements across the country. To do so, the household registration systems connect urban and rural populations to different sets of public service resources within their administrative region, including health care, education, and housing. The Pearl River Delta, where Guangzhou is located, has the largest urban agglomeration globally based on the total residential population (UN Habitat 2016). The city's gross domestic product (GDP) increased 56 folds, from approximately

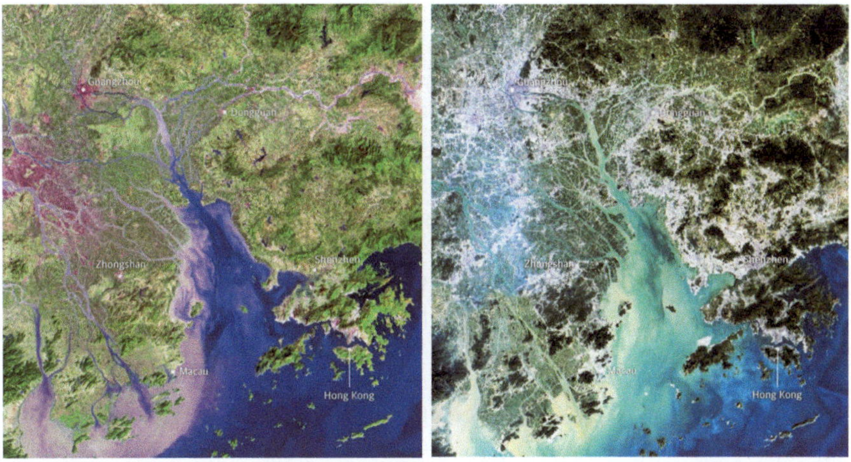

Fig. 5.4 Pearl River Delta in 1979 (left) and in 2000 (right), Comparing to the previous image, vast swathes of farmland have turned from green to grey, and huge new areas of land reclaimed from the sea. Satellite image 1979 and 2000 © Landsat/Nasa/UIG/Getty images

Fig. 5.5 Guangzhou South Avenue looking towards the river in 1991 and 2015 Left: 1991 © Zeng Zhi; right: 2015 © Alex Hofford

32 billion (CNY, ~ 4 billion Euro) in 1990 to 1 trillion and 810 billion (CNY, ~226 billion Euro) in 2015. All three districts selected for this study: Haizhu, Tianhe, and Panyu, became Guangzhou's administrative districts by 2000. Urban development land availability decreased drastically in these districts over the last few decades. Among them, Haizhu and Tianhe Districts are significantly smaller in land area and their residential population density is significantly larger than other districts (also see Table 5.1). These two indicators are of great significance from an urban land economy perspective. While Panyu District has mostly industrial land even though the land area is relatively large. Since Urban green spaces is regarded as a part of urban development land in the Chinese land-use system, the total amount of usable urban land and how it is being used are of great significance to Urban green spaces planning and governance.

Table 5.1 Land area, population density, and Urban green spaces size of each district in Guangzhou in 2016

	Panyu	Haizhu	Tianhe	Guangzhou total
Land area (Km^2)	529.94	90.40	96.33	**7,434.40**
Residential population (million)	1,641.1	1,637.9	1,631.0	**14,043.5**
Population registered as urban households (million)	886.5	1,022.6	867.7	**8,704.9**
Density of residential population (per Km^2)	3,097	18,118	16,931	**1,889**
Density of population registered as urban households (per km^2)	1,673	11,312	9,008	**1,171**
Size of Urban green spaces (ha)	1,731.84	342.96	2,605	**15,750.93**
Size of Urban green spaces per capita (m^2)/ residential population	10.55	11.06	15.97	**16.5**

Data source Guangzhou Municipal Bureau of Statistics and Bureau of Forestry and Landscaping

5.1.2 Climate Change Impacts

China's geographic and climatic conditions and population distribution determine that the country is highly susceptible to climate change impacts. Although China's climate is conducive to the cultivation of agricultural products, the disadvantage is that the country has frequent disastrous weather, among which droughts, floods, cold waves, and typhoons are the main hazards that cast negative impacts on the population often. In summer and autumn, tropical cyclones or typhoons frequently hit the southeast coast of China. During the autumn and winter seasons in China, cold air from Mongolia and Siberia sweeps southward, sharply decreasing the temperature and forming cold waves, frosts, and strong winds, which impact human health negatively and can jeopardise agricultural production.[3]

Climate change is evident in China. From 1901 to 2017, the average surface temperature in China has become 1.21° higher than the preindustrial level, compared to the global average of about 1.1 °C (National Climate Centre 2018). The sea level in coastal China rose around 122 mm from 1980 to 2017 (China Meteorological Administration 2018). Looking forward, the probabilities of continued sea-level rise and the consequential extreme events, such as tropical cyclones, storm surge, and coastal erosion, are very likely to increase significantly over the twenty-first century and the coasts of China are among the highly vulnerable coastal regions (Feng and Tsimplis 2014; Hallegatte et al. 2013b; Yang et al. 2017). Climate change has caused the 400 mm isoline of mean annual precipitation to shift around, but more importantly has made extreme climate events such as typhoons, storms, and floods, more intense and more frequent in the east side of the Heihe-Tengchong Line (see Fig. 5.1). Notably, one recent study projects that 37 million people in China will be exposed to

[3] State Council of China Homepage > basic conditions of the country (in Chinese: 首页>国情). Available at: http://www.gov.cn/guoqing/index.htm.

sea-level rise and related consequences, accounting for 53% of the global exposure (Muis et al. 2016). So far, climatic hazards have costed an average $76 billion US dollars in China and affected around one-third of agricultural land (Asian Development Bank 2015). While poverty rate has been drastically reduced over the last few decades (Hu et al. 2003), many urban areas in China are not decently equipped for natural hazards due to infrastructural creation failing to catch up with the speed of urbanisation (Xiao et al. 2018; Zuo 2014).

As a low-lying coastal city in the Pearl River Delta, Guangzhou is easily affected by rising sea levels. Its unique Asian subtropical monsoon climate also means that millions of people in the city increasingly need to withstand the direct threats of tropical cyclones, heavy rainfall, floods, high temperature and heatwaves, haze, thunder and lightning, and sea-level rise. Since the New Millennium, the threat of intensified heatwaves on the city became particularly evident. Regular heatwaves throughout the summers in Guangzhou have intensified in the last 60 years, both in the number and duration of heatwaves (see Fig. 5.6). From 2001 to 2010, the number and time of heatwaves in Guangzhou have dramatically increased. Such results from climate change and the rising heat-island effect induce significant health impacts ranging from insomnia, fatigue, and clinical exacerbation to increasing malaria transmission potential to death from heatstroke (Du et al. 2013).

The China-UK Global Health Support Programme's research on Guangzhou finds that the higher intensity and time length a heatwave possesses, the higher the mortality risks become. According to the Guangdong Provincial Centre for Disease Control and Prevention (GPCDCP), each degree of temperature rise in Guangzhou has increased

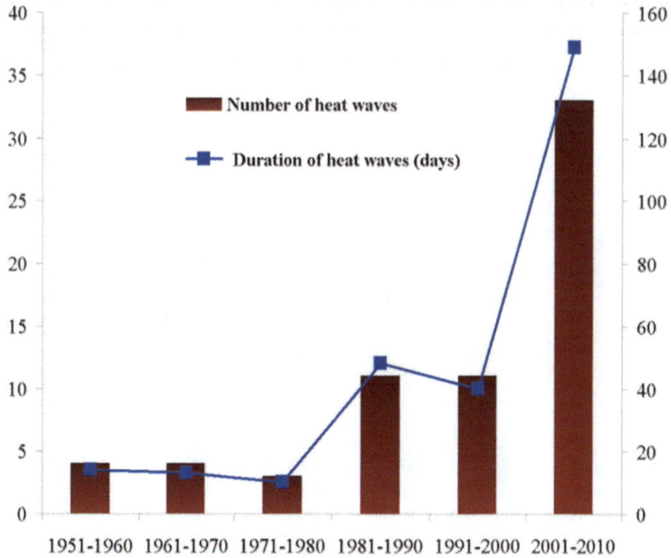

Fig. 5.6 Decade trend of heatwave during the last 60 years in Guangzhou. *Source* The Adapting to climate change in China project (ACCC 2012)

3.7% mortality rate for the elderly population (≥65 years old) (Liu 2020). Furthermore, due to the stagnant groundwater and hot and humid weather, dengue fever is a common concern in post-typhoon periods. Climate change, with its intensification effect on the length and intensity of typhoons, as well as rainfalls, has demonstrated significant impacts on dengue in Guangzhou (Xiang et al. 2017).

Besides, tropical cyclones affecting the PRD Region in the last decade became more intensive and brought heavier rainfalls (see Fig. 5.7). The intensity of typhoons generated in the western North Pacific and the consequential storm surges have increased due to warmer ocean surface temperatures and decreased salinity (Balaguru et al. 2016; IPCC 2014; Nakamura et al. 2016). The typhoons that demonstrated destructive power in Guangzhou and the neighbouring provinces between 2013 and 2017 have caused an estimated 2000 deaths (State Council of China 2014a). In addition, the increased intensities of the strongest tropical cyclones and the rising sea levels globally continuously impact the social, economic, and environmental determinants of mental health (Elsner et al. 2008; Fritze et al. 2008; Walsh et al. 2014). These extreme events may lead to serious mental health problems, such as fear and stress over exposure to greater disaster risks and post-disaster trauma-related mental illness, possibly including increased suicide mortality (Berry et al. 2010). The impacts of climate change on mental health merit further investigation.

Guangzhou has long been ranked as one of the most vulnerable cities in the world facing climate change, as shown in various climate modelling results. The city was ranked as the second most vulnerable city among the top ten, only after Mumbai, in terms of exposed population (including all environmental and socio-economic factors) for present-day conditions (2005). The impacts of climate change on human health and well-being are concrete, complex, and intricate. These impacts tend to bring forth more significant damages for physically vulnerable groups. Vulnerable urban populations, e.g., rural migrants provided with basic housing conditions such

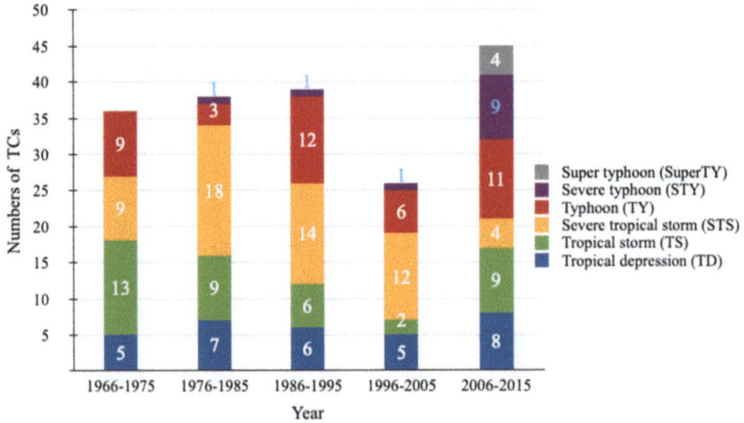

Fig. 5.7 Numbers of landfalling tropical cyclones affecting the Pearl River Delta region from 1966 to 2015. *Data source* HKO's tropical Cyclone dataset 2016

as pregentrified urban villages and peri-urban areas and the urban poor who possess weaker financial means, are especially vulnerable to the adverse health impacts of climate change. The fact that climate change-related mortality is on the rise could indicate a lack of knowledge on how climate change impacts health and which actions can be taken to mitigate the impact.

Two-thirds of the world's largest cities are in coastal deltas; Guangzhou is one of them. Due to the low-lying geographical feature, these cities are particularly vulnerable to climate change impacts such as rising sea levels. A global chemistry-climate model projected that under the high greenhouse gas scenario RCP8.5, premature mortality from climate change-attributed air pollution increases in all regions, with the most significant rise in India and East Asia, which includes Guangzhou (Silva et al. 2017; Du et al. 2013). Research on vulnerability variation from coastal regions to inland in Guangzhou indicated that climate change is likely to bring stronger heatwaves in the future, thereby increasing heatwave-related illnesses and deaths (Du et al. 2013; Liu et al. 2013; Zhu et al. 2014). The same China-UK research also finds that heatwaves affect health more strongly in Chinese urban areas where air conditioners are less commonly seen, and disposable income is lower per capita. In addition, recent research also points to the intensification of cold spells due to climate change and that the cold spells related to non-accidental deaths in Guangzhou will increase (Sun et al. 2018). By the 2070s, Guangzhou is projected to still be among the world's top ten most vulnerable cities, irrespective of the changes, and the social and economic loss could be catastrophic (C40 Cities Climate Leadership Group, n.d.; Nicholls et al. 2007).

5.2 Institutional Contexts of China

5.2.1 Contemporary Development Philosophy Before Ecological Civilisation

5.2.1.1 Socialist Market Economy with Chinese Characteristics

Scholars consider the period between the founding of the Chinese socialist state in 1949 and the economic reform in 1979 as the modern political period characterised by the communist identity, state-controlled resource allocation, and the ethical norm of prioritising the common good over individual good (Dallmayr et al. 2012; Li 2018). The period since the economic reform is generally considered as the contemporary political period marked by the state-oriented market economy, or, as formally named, socialist market economy with Chinese characteristics (in Chinese: 中国特色社会主义市场经济 zhōng guó tè sè shè huì zhǔ yì shì chǎng jīng jì) (Cheng and Chen 2018). With the economic Reform and Opening Up since 1979, particularly since the early 1990s, China underwent a rapid process of marketisation of goods and services by privatising property rights orchestrated by the state. The process is accompanied by the transition of governance and regulation based primarily on political means (or ideological regulation) to legal means and economic means. State intervention in society, in general, saw a decrease. The goal of the transition in the Chinese political and economic transition is to develop a civil society through economic development, industrialization, and urbanisation, while keeping the emphasis on social harmony and the performance of social roles that the leadership promotes within the country's contemporary social, economic, historical, and cultural context (Li 2018).

The lack of former references and experience challenged the central government in leading the path to marketization and industrialization, followed by the local governments. Problems frequently arose. For example, the privatisation process, primarily when conducted by local governments, was not standardised, resulting in the devaluation and sometimes even corruption of state assets. Besides, the central government's vision and agendas often did not align well with the needs or interests of the local developments, resulting in coordination problems. Furthermore, the economic dynamics fostered in contemporary Chinese society determine that all levels of government provide policy support and gain economic benefits. Local governments have increasingly valued economic benefits, gradually prioritising projects with high and fast economic benefits. On this basis, the state has gradually prioritised economic efficacy over distributive justice, not only economic justice but also social and environmental justice.

5.2.1.2 Scientific Outlook on Development

After summarising the above problems and experience in the Reform and Opening up, in 2003, the former president Hu Jintao stated at the 16th Party Central Committee Session that it is necessary to "adhere to a *people-centred principle* and establish a comprehensive, coordinated and sustainable development concept." In 2007, the 17th National Congress defined *Scientific Outlook on Development* (科学发展观, kē xué fā zhǎn guān) as *sustainable development* that is *people-centred*. This was the first time the term *people-centred* was made formally associated with the notion of *sustainable development*. The goal is to address the high cost of the "pollute first, control later" approach taken since the Reform and Opening up and to achieve comprehensive, coordinated, and sustainable development through coordinated and comprehensive planning. The objective is to "pursue comprehensive progress both at individual and societal levels based on economic development, develop socialist economic, political, and spiritual civilization through coordinated means, and insist on achieving human-nature harmony in natural resources consumption, ultimately achieve sustainable economic and social development." (Literature Research Office of the Central Committee of the Communist Party of China 2005: 483). Furthermore, in 2006, the national government made it a strategic task to build a "resource-conserving, environment-friendly society" in medium- to long-term socio-economic planning.

5.2.2 Ecological Civilisation

This section introduces *Ecological Civilisation*, China's current most ambitious national environmental governance framework, in four parts: (1) the emergence of the concept, (2) the Integrated Reform Plan of a more comprehensive environmental governance system, (3) experiments and development of novel concepts including Gross Ecosystem Product (GEP) and Ecological Asset (EA) for fundamental value change, (4) *Ecological Civilisation* operationalised as *people-centred development* in the urban context, and 5) challenges of insufficient understanding of the current institutional reform.

(1) **China's emerging national environmental governance framework**

In 2007, the concept of *Ecological Civilisation* (生态文明 sheng tài wén míng) was introduced into the National Party Congress Report as "a **future-oriented guiding principle** based on the perception of the extremely high price we have paid for our economic miracle" and as a strategic goal to establish a more comprehensive environmental governance institutional framework (China Daily 2007). The concept was proposed as a core national development strategy in 2012 (Pan 2015). *Ecological Civilisation,* according to senior Chinese development officials, emphasises respect for nature and protection of people's rights and aims to achieve ecological and social justice, improve efficiency in economic production and balance of regional differences, and secure quality life for the people (Pan 2015).

The concept of *Ecological Civilisation* builds upon the realisation that gradually came through about the high ecological footprint and unsustainability of the Western Industrialisation development pathway, after following this model for more than three decades. China has achieved great economic success but, in the meantime, experienced severe environment deterioration, increasing inequality, and loss of critical Ecological Assets. The Western development model since the Industrial Revolution has been increasingly referred to by scholars as the "develop (the economy) first and treat (the environment) later" model (先发展后治理 xiān fā zhǎn hòu zhì lǐ). In fact, as early as in the 1920s, British historian Bertrand Russell (2009) argued that industrialism is a threat to human freedom, especially mental freedom because of its fundamentally exclusive nature (private property), uninspiring form (mechanistic outlook), and egoistic political purpose (nationalism). Entering the 1960s, a series of environmental incidents prompt a broad societal reflection on economic growth and environmental movement. Rachel Carson (2000) described a world possibly without birds, bees, and butterflies due to the extensive use of pesticides and artificial fertilisers. Donella Meadows (1972) inquired about the quantitative constraint of global resources and the system of production on human socio-economic activities and emphasised the criticality of redirecting society towards global economic and ecological equilibrium instead of growth.

In addition, the Chinese traditional thoughts of nature, such as the "human-nature oneness" from Taoism (also see Sect. 2.1.2 Ecological notion and Governance philosophy in China), has gradually become the philosophical foundation which formed the principle of *Ecological Civilisation.* In China, agricultural economist Ye Qianji first introduced the concept from the former Soviet Union in 1987 and further conceptualised it based on the traditional Taoist thinking on the human-nature relationship (Xu 2009). For instance, the Chinese sage Laozi wrote in the Tao *Te Ching*, the sixth-century BC Chinese classical text: *"Humans produce and reproduce according to (the ability of) land, land produces according to the weather, the weather serves the land according to Tao, and Tao is subjected to Nature's grace* (in Chinese: 人法地, 地法天, 天法道, 道法自然; rén fă de, dì fă tiān, tiān fă dào, dào fă zìrán)." "*Tao* (道)" in Chinese has the literal meaning of "way, methods, or principle." The concept of *Ecological Civilisation* is built upon the idea that human behaviours need to follow and respect nature's capacity (Pan 2015).

In short, the concept of *Ecological Civilisation* contains explicit value prospects and institutional goals necessary for building a new political philosophy in China that defines a development paradigm highlighting fairness, efficiency, harmony, and human development (Pan 2015). On value, the phrase "*lucid waters and lush mountains are invaluable assets*" (in Chinese: 绿水青山就是金山银山, lǜ shuǐ qīng shān jiù shì jīn shān yín shān) proposed by the current President Xi Jinping has been widely promoted as an interpretation of *Ecological Civilisation*—a value of respect for nature, co-existence, and co-prosperity with nature. On institutions, it seeks to establish a system based on regulations, rules, and economic incentives to improve nature conservation and protection, production of higher efficiency, and distribution of goods and services of better fairness.

(2) **An Integrated Reform Plan of a more comprehensive environmental governance system**

Initially, the concept of *Ecological Civilisation* was criticised as vague and little operationalisable (Marinelli 2018). After defining the mission and main task of the future-oriented guiding principle *Ecological Civilisation*, the State Council issued the ***Integrated Reform Plan for Promoting Ecological Progress*** (in Chinese: 生态文明体制改革总体方案; shēng tài wén míng tǐ zhì gǎi gé zǒng tǐ fang àn) in 2015 to put forward a more systematic, comprehensive, and better-coordinated institutional framework. As showcased in Table 5.2, this Integrated Reform Plan established explicitly eight major working areas consisting of 47 working items of environmental governance (State Council of China 2015).

These working areas and items proposed above were deemed innovative, visionary, and ambitious (Li [ed.] 2018), therefore required a suitable and effective government system for implementation. At that time, three main problems were identified in a national government performance evaluation: (1) disarrangement of positions, powers, and responsibilities of some government agencies including central and local ones, leading to (2) inefficient performances, and (3) the lack of effective power checking, monitoring, and supervision mechanism (Wei 2018). To improve the government system's suitability and effectiveness and establish the above-mentioned

Table 5.2 Major working areas and working items listed in the Integrated Reform Plan for Promoting Ecological Progress

	Working areas	Working items
1	Property rights of natural resources	1. Establish a unified system for determining and registering property ownership rights 2. Establish a system of property rights for natural resources with defined rights and responsibilities 3. Improve the state natural resource assets management system 4. Establish a system for exercising property ownership rights at different administrative levels 5. Launch demonstration projects for determining and registering watershed and wetland ownership rights
2	Land space development and protection	6. Improve the zoning system of functional zones 7. Improve the regulatory system of land space use 8. Establish a national park system 9. Improve the system of natural resources regulation
3	Spatial planning	10. Formulate spatial plans 11. Promote integrated planning at municipal and county levels 12. Develop new approaches for formulating spatial plans at municipal and county levels
4	Total resource management and conservation	13. Reinforce systems for farmland protection and sustainable land-use 14. Strengthen a system for water resources management 15. Establish a system for total energy consumption management and conservation 16. Establish a system for old-growth forest protection 17. Establish a system for grassland protection 18. Establish a system for wetland protection 19. Establish a system for deserted land close-off and protection 20. Improve the system of marine resources development and protection 21. Improve the system of mineral resources development and utilisation 22. Improve the system of resource recycling
5	Payment-based natural resource consumption and ecological compensation	23. Accelerate the reform for natural resource products pricing 24. Improve the payment-based system for land-use 25. Improve the payment-based system for mineral resource use 26. Improve the payment-based system for use of oceanic spaces and offshore islands 27. Accelerate the reform of resource and environmental taxes and fees 28. Improve the ecological compensation system 29. Improve the mechanism for utilising ecological protection and restoration funds 30. Create a system for farmland, grasslands, rivers, and lakes recuperation

(continued)

Table 5.2 (continued)

	Working areas	Working items
6	Environmental governance system	31. Improve the pollutant emissions permit system 32. Establish a mechanism for regional cooperation in pollution prevention and control 33. Establish systems and mechanisms for rural environmental governance 34. Improve the systems of environmental information public disclosure 35. Strengthen the implementation of ecological and environmental damage compensation systems 36. Improve the environmental protection management system
7	Market system for environmental governance and ecological conservation	37. Foster market entities for environmental governance and ecological conservation 38. Promote the energy use and carbon emission permits trading systems 39. Promote the pollution permit trading system 40. Promoting the water rights trading system 41. Establish a green finance system 42. Establish a unified system for green products
8	Environmental governance performance evaluation and accountability system	43. Establish an Ecological Civilisation targets system 44. Establish monitoring and early warning mechanisms for environmental and resource carrying capacity 45. Create natural resource assets' balance sheets 46. Audit outgoing officials' management of natural resource assets 47. Establish a lifelong accountability system for ecological and environmental damage

Data source State Council of China (2015)

Ecological Civilisation institutional framework, the State Council issued the *Decision of the Central Committee of the Communist Party of China on Some Major Issues Concerning Comprehensively Deepening the Reform* (in Chinese: 中共中央关于深化党和国家机构改革的决定; zhōng gòng zhōng yāng guān yú shēn huà dǎng hé guó jiā jī gòu gǎi gé de jué dìng) in 2018. The *Decision* gave instructions on how to **provide institutional guarantee for building *Ecological Civilisation*, namely, by "reforming the management and protection systems of natural resources and ecological environments, and building an environmental governance system led by state, primarily participated by market, and co-participated by societal organisations and the public**" (State Council of China 2018). An immediate subsequent action of this *Decision* was the creation of two brand new ministries: the Ministry of Ecology and Environment and the Ministry of Natural Resources, concentrating the scattered responsibilities of different government departments on environmental governance (see Fig. 5.8).

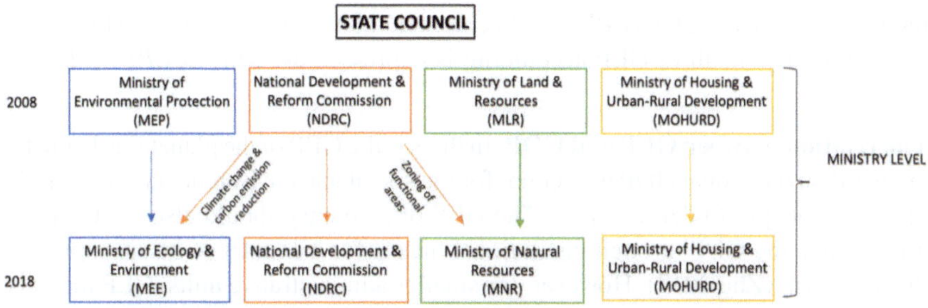

Fig. 5.8 Institutional reform at the national level in the Chinese central government in 2018

(3) Examples of experiments and development of novel concepts: gross ecosystem product (GEP) and ecological Asset (EA) for fundamental value change

Human societies and the ecological environment that they depend on for survival and development constitute complex social-ecological systems. Environmental and especially ecological economists have long advocated for accounting ecosystem services that human societies and economic development have benefited from, with the support of less economy-centric references such as the Human Development Index (HDI) and Gross National Happiness (GNH). The *Integrated Reform Plan* introduced above-proposed establishing diverse mechanisms for assessing the ecological dependence for ecological compensation, restoration, and conservation based on quantitative evidence (Ouyang et al. 2018). Gross Ecosystem Product (GEP) is the total economic value of ecosystem products and services for human welfare and sustainable development in a geographical area with one year as a time unit for analysis (Ouyang et al. 2013). Ecological Asset (EA) refers to various ecosystems that produce and provide ecosystem products and services, including natural and artificial ecosystems based on natural ecological processes such as urban green spaces. The GEP and EA accounting methods and experiments are also expected to lead to the development of other ecological governance tools, such as the Ecological Assets' balance sheets and the auditing system for outgoing officials' performance on natural resource management.

The work of establishing the GEP accounting system under *Ecological Civilisation* in China has been based on many concluded or ongoing international ecological environment protection work, including the UN-led Millennium Ecosystem Assessment, the Intergovernmental Science-Policy Platform on Biodiversity and Ecosystem Services (IPBES), the UN System of Environmental-Economic Accounting (SEEA) integrated economic and environmental data framework, the Economics of Ecosystems and Biodiversity (TEEB) initiative, and the World Bank Wealth Accounting and the Valuation of Ecosystem Services (WAVES) global partnership programme.

GEP accounting includes the total direct- and indirect-use values of ecosystem goods and services, including ecosystem provision value (EPV), ecological regulation services value (ERV), and ecological culture services value (ECV) (see Table 5.3). The math of GEP accounting is expressed as: $GEP = EPV + ERV + ECV$.

The relation between GEP and GDP: In theory, the GEP of the planet Earth must be greater than the global GDP to account for the social and environmental externalities that the concept of GDP failed to. The GEP of a country should also be defined as greater than the GDP through institutional means for a country to be sustainable in the long term (Zhu 2017). However, for smaller administrative units, GEP may also be smaller than GDP. This is often the case for cities with a service-based economy and high population density (e.g., Shanghai, London, and New York), which depend on imports of ecological resources.

(4) ***Ecological Civilisation* operationalised as *people-centred development* in urban context**

Today, *Ecological Civilisation* has evolved into a comprehensive institutional framework encompassing guidelines, indicators, assessment methods, standards that orientate China's environmental policy-making and governance from central to local levels across land-use, natural resource conservation, ecological protection and restoration, urban planning, pollution emission control, and many more. One of the main concepts of *Ecological Civilisation* is "people-centred development." As proposed by former president Hu Jintao, adhering to people-centred development means "to achieve the goal of comprehensive development of people, to seek and promote development from the fundamental interests of the people, to continuously meet the growing material and cultural needs and to effectively protect the economic, political and cultural rights and interests of the people" (People's Daily 2008). The people-centred development principle emphasises economic growth to achieve the end goal of fulfilling the material and cultural needs of people (People's Daily 2008).

The people-centred development principle is applied in the National New Urbanisation Plan (2014–2020) to address the country's widespread urban problems,

Table 5.3 Types of values of ecosystem goods and services considered in the GEP accounting

Types of values	Examples
Ecosystem provision value (EPV)	E.g., agricultural, forestry, livestock, and fisheries products, water resources, energy, and other tangible products that enter the market
Ecological regulation services value (ERV)	E.g., water and soil conservation, water purification, flood storage, sand fixation, air purification, carbon sequestration, and climate regulation
Ecological culture services value (ECV)	E.g., recreation, mental and physical health, tourism, aesthetic appreciation and inspiration for culture, art and design, spiritual experience, and sense of place

Source Ouyang et al. (2013, 2018), the Economics of Ecosystems and Biodiversity (TEEB)

including the marginalisation of urban migrants and environmental pollution, and to transform its land-centred urbanisation practices (Chen et al. 2016). This New Urbanisation Plan elaborated measures on integrating rural to urban household registrations, improving public transport, housing, utility infrastructure, healthcare and education services, institutionalising ecological and environmental protection, and strengthening and developing new approaches to social governance (State Council of China 2014c).

(5) **Insufficient understanding of the current institutional reform under *Ecological Civilisation***

Despite the strong political determination of the Chinese central government to conduct institutional reform and create a new environmental governance framework, the process has not always been smooth. A recent large-scale survey conducted by the National Development Research Centre for local government officials and policy experts shows mediocre knowledge rates on the content of the *Integrated Reform Plan*. Only about half of the survey participants knew all the eight major working areas, and few recognised all the 47 working items. In general, local policymakers tend to have better knowledge of the more coercive working items and the ones with higher economic sensitivities, such as the auditing system for outgoing officials' natural resource management performance and the lifelong accountability system for ecological and environmental damage, and the ecological compensation system.

5.2.3 History of Environmental Protection in China

Regardless of political philosophy, the development and modernisation of a society require political ambitions and strategies to organise resource distribution properly. Development and modernisation across different societies share the common elements of sovereign responsibility, resource distribution, and a range of political means and ends. The Chinese government has followed the international environmental agendas since attending the first UN Conference on the Human Environment in Stockholm in 1972. Since then, the central government has undertaken major reforms in the national environmental governance system, all towards strengthening the structure and position of the national environmental agency and building a more comprehensive institutional framework combining regulatory and economic instruments. For example, the State Council passed the *Environmental Protection Law*, the *Law on the Prevention and Control of Atmospheric Pollution*, and the *Environmental Impact Assessment law* in 1979, in 1987, and in 2002 respectively. These laws contained regulatory instruments such as environmental quality and emission standards, discharge permit system, as well as economic instruments such as pollution charges specific to both the types and volume of pollutants (OECD 2006). Central government´s determination for environmental regulation was particularly evident between the mid-1990s and the New Millennium, explicating environmental governance goals through various action plans and enacting new environmental laws frequently, including the Five-Year Social and Economic Development Plans (FYPs) and the correspondent Environmental Plans with industry-specific, time-bound targets. The State Council launched *China's Agenda 21* in 1994 (State Council of China 1994), the first white paper focusing on the nexus between its population growth, environment potential, and visions of development in the twenty-first century.

In 1996, *Sustainable Development* was elevated as a national strategy to guide the large-scale, rapid urbanisation process of the country. In 2003, the former president Hu Jintao proposed the *Scientific Outlook on Development* (科学发展观kē xué fā zhǎn guān) which can be summarised succinctly as people-centred, comprehensive, coordinated, and sustainable development, or *people-centred sustainable development* (Pan 2015) (also see Sect. 5.2.1 Contemporary development philosophy before *Ecological Civilisation*). In 2003, the central government adopted a form of *Strategic Environmental Assessment (SEA)* in order to insert more valuable notions of sustainable development into the country's macro socio-economic development plans (Zhu and Ru 2008). The central government further proposed environmentally relevant concepts such as the "Resource-Conserving, Environment-Friendly Society" in 2006 and formalised it as a strategic task in medium- to long-term socio-economic planning. Major institutional reforms of China's environmental governance system led by the state prior to *Ecological Civilisation* are shown in Fig. 5.9.

Prior to the State Environmental Protection Agency (SEPA) being upgraded to the Ministry of Environmental Protection (MEP) in 2008, China's environmental management system had limited authority and capacity to be effective and efficient in policy implementation (OECD 2006). That was due to the provincial and local branches of the SEPA being coordinated and directed more so by local governments than by the SEPA hierarchically from Beijing. While SEPA led a strong environmental agenda at the central level, local governments often still had alternative

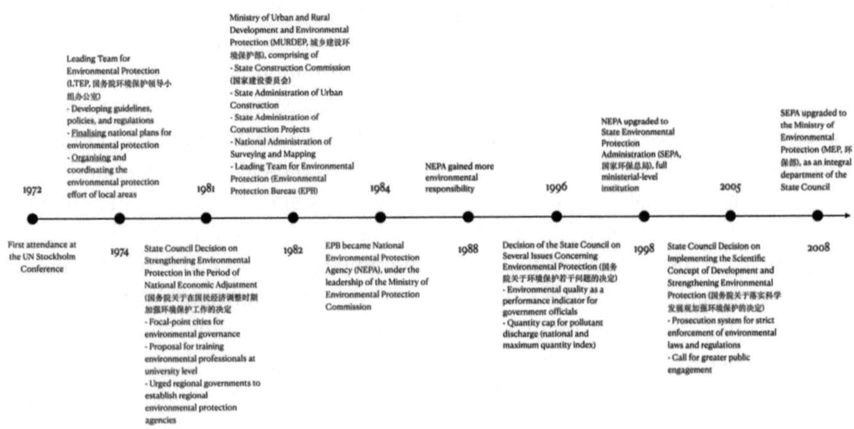

Fig. 5.9 Major institutional reforms of China's environmental governance system led by the state prior to *Ecological Civilisation. Data source* Wang (2019), MEE, OECD (2006), State Council (1996, 2005, 2011)

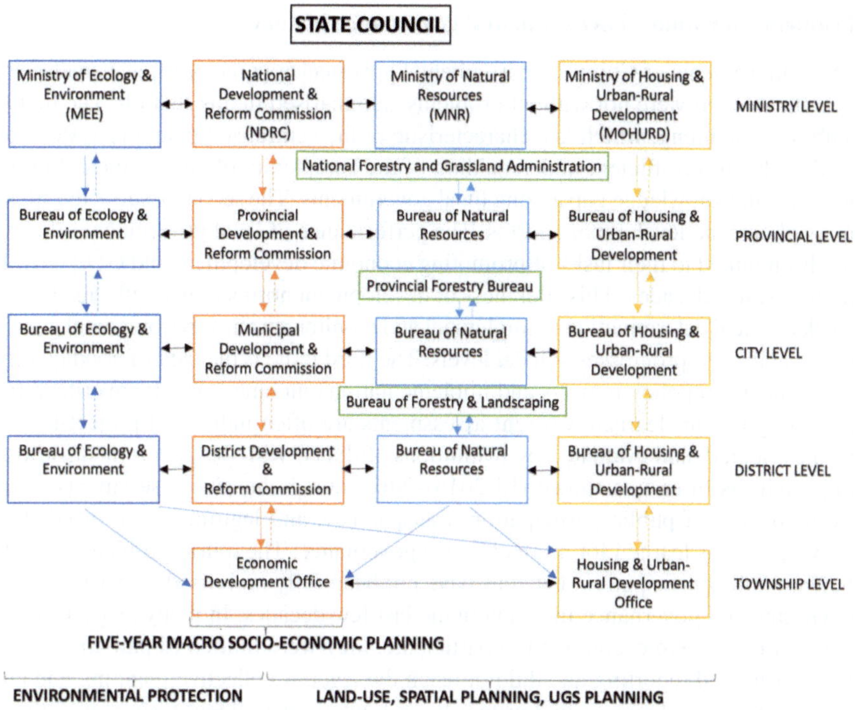

Fig. 5.10 Current five-level administrative system of environmental protection, land-use, and spatial (and Urban green spaces) planning

priorities (e.g., economic development), leading to weak financial and institutional support for implementing environmental policies at local levels.

For this reason, the State Council has gradually made it a work focus to improve the institutions for better local environmental governance, so that environmental agendas set at the central level could be implemented properly. In 2011, the State Council issued the *State Council Opinion Regarding Strengthening Key Environmental Protection Work* (in Chinese: 国务院关于加强环境保护重点工作的意见; guó wù yuàn guān yú jiā qiáng huán jìng bǎo hù zhòng diǎn gong zuò de yì jiàn), urging for (1) improving the level of supervision and management in environmental protection, (2) focusing on resolving prominent environmental problems that jeopardise public health, and (3) reforming and innovating the institutional framework and mechanism of environmental protection. Furthermore, the *Integrated Reform Plan for Promoting Ecological Progress* (in Chinese: 生态文明体制改革总体方案; sheng tài wén míng tǐ zhì gǎi gé zǒng tǐ fang àn) (introduced in Sect. 5.2.2 *Ecological Civilisation*) outlined more actions to establish a more systematic, holistic, and better-coordinated systems for environmental governance, including the creation of two brand new ministries exclusively dedicated to environmental and natural resource governance.

Problems in China's environmental governance history

The achievements of China's environmental governance in the past few decades are applaudable but were not without problems, summarised in this section. The first is ambitious planning, which is a characteristic of the centralised planning system, as well as due to insufficient understanding of the complexity of environmental problems and the knowledge gap among local governments. The second issue is the stickiness of current development norms. The performance of local government officials has been linked to their tasks of promoting economic development and GDP growth in the last few decades. This stickiness of development norms is also reflected by the weak political determination to implement and monitor current environmental governance plans and progresses at local levels. The third issue is the lack of good quality assessments on policy impacts and implementation outcomes, which are critical for providing lessons learned. Current assessments are often individual project-based, result-oriented, and evaluated by institutions affiliated with the government, whose objectivity is often questioned (Li 2016). More comprehensive assessments, such as the degree of public participation, transparency, and legitimacy, are needed to provide lessons learned for more effective governance. The fourth problem is insufficient public participation. Citizens have not been fully mobilised to participate in environmental governance in China in the last few decades. In many projects, even when citizens were encouraged to participate, they were limited to participating in monitoring and reporting and did not reach the level of collective planning and co-implementation. Having relied primarily on limited government resources without sufficient public participation might have been another reason why environmental governance in China still has room for improvement.

5.2.4 Important Plans on Environmental Health and Climate Change

This section introduces six major national plans on environmental health and climate change: (1) the *National Environmental Health Action Plan (2007–2015),* (2) the *National Strategy for Climate Change Adaptation*, (3) the *Action Plan on Prevention and Control of Air Pollution 2013*, (4) *National Plan on Climate Change (2014–2020)*, (5) *Action Plan for Urban Climate Change Adaptation*, and (6) *Healthy China 2030*. These national plans have direct and explicit reasonings for the significance of Urban green spaces and commands for planning Urban green spaces as actions for respective environmental health and climate change adaptation purposes.

1. **The *National Environmental Health Action Plan (2007–2015)*** is the first national guiding principle addressing the environment-health nexus and calling for action. It is coordinated by the Ministry of Health and sixteen other ministerial agencies. The plan noted that environmental degradation, the destruction of ecological balance, and harm to human health have "increasingly become the

key factors constraining economic growth and affecting the harmonious development of society. The plan strives to solve major environmental problems that endanger people's health and further promote sustainable socio-economic development. It has defined it as both a "basic requirement and an important task" for establishing "people-oriented governance" and building a "harmonious socialist society (State Council of China 2007: 3)." This plan highlighted the significance of resolving major environmental health problems through environmental health assessments—namely, the impacts of climate change on human health and coping strategies. The effects of climate change on human health referred to in this plan include extreme weather and climate events such as heatwaves, storm floods, dust storms, droughts, floods, and other climate change-induced or intensified diseases. Besides, the plan also emphasised the importance of establishing early warning systems, emergency response plans, related methods and technologies, and effectiveness evaluations of these interventions.

2. **The *National Strategy for Climate Change Adaptation*** has debuted during COP19 in Warsaw in 2013. The strategy was the first and one of the country's most critical sets of climate change adaptation policies. It outlines a systematic transition framework of work focus from climate change assessment to adaptation policy-making, and it serves as guidance for cities to make climate change adaptation plans at their respective level (State Council 2013). The strategy has defined the goal of protecting health from increasing risks of occurrence and spread of many diseases, especially temperature-sensitive illnesses and post-disaster infectious diseases. Specifically, the strategy highlights human health as one of the seven important tasks to be implemented by 2020. This includes: (1) establishing climate change-health monitoring, assessment, and early warning systems; (2) improving the control system of infectious diseases; (3) wide spreading the knowledge and skills of health protection facing climate change; and (4) enhancing health supervision and law enforcement. Climate change researchers also back the necessity of establishing early warning and monitoring systems (Du et al. 2013).
3. **The *Action Plan on Prevention and Control of Air Pollution 2013*** is an outcome of the seventh National Conference on Environmental Protection 2011. The conference framed decent environmental quality as a public good that governments should provide, which requires resolving major environmental problems that jeopardise human health and development. Thus, the action plan sets human health as both the entry point and fundamental interest and has laid out ambitious, specific targets from the national level to reduce ambient air pollution and improve air quality nationwide by the end of 2017. It aimed to reduce PM10 by at least 10% relative to 2013 levels across all urban areas. For Beijing, the target was to reduce PM2.5 by 34% from the 2013 level. The most heavily polluted and most densely populated urban areas in the country: Beijing-Tianjin-Hebei, Pearl River Delta, and Yangtze River Delta, received specific targets to reduce PM2.5 by 25%, 20%, and 15%, respectively.
4. ***National Plan on Climate Change (2014–2020)***. Several incidents prompted China's increasingly affirmative gesture on environmental and climate change

governance. China overtook the USA and became the biggest CO2 emitter in 2007 and gained international exposure from hosting the International Olympic Games in 2008. Air pollution has become a genuine concern of citizens as more evidence surfaced, indicating negative health consequences. There is an evident change in the central government's political gesture towards the challenges of climate change impacts. As shown in the *National Plan on Climate Change (2014–2020)*, climate change was deemed "crucial" in influencing the overall socio-economic development, energy security, food, the ecological environment, and human health. Besides, the plan frames "actively responding to climate change" as an inherent requirement for achieving *Ecological Civilisation* and a significant opportunity for transforming the current economic development model. The plan further frames it as an "international obligation" of the country as a "responsible powerful nation."

5. **The Action Plan for Urban Climate Change Adaptation** focuses on the role of cities in climate change adaptation, as cities host a significant quantity of social and economic activities. The urban population in Guangzhou is exposed to high risks of severe climate change impacts, mainly due to its high density. The *Action Plan for Urban Adaptation to Climate Change* (State Council of China 2016b) was co-carried out by the National Development and Reform Commission (NDRC) and the Ministry of Housing and Urban–Rural Development (MOHURD) in 2014. The action plan establishes that "urban adaptation to climate change is about the people's vital interests." Besides, it supports cities to "actively promote" climate change adaptation by addressing public health, among other aspects (depending on each city's specific climate change challenges), in their pilot work programmes. The plan emphasises that cities taking action to adapt to climate change are an essential contribution to achieving *Ecological Civilisation.* The action plan requires municipalities to make "more comprehensive and forward-thinking" policies for climate change adaptation. For Guangzhou, this implies making policies incorporating updated scientific evidence to manage the impacts of storms and floods effectively.
6. **Healthy China 2030.**The State Council endorsed the Healthy China 2030 Plan in 2015, which entered into effect in 2016. This is China's first national medium- to long-term action plan strategically focusing on health. It aims to promote healthy lifestyles, improve health services and the health industry. Besides, the plan outlines the vision of a sustainable health system to provide essential health services to every citizen by 2020. The plan defines health as a "basic condition for socio-economic development." It states that improvement in national health governance is a "proactive contribution to global health governance and fulfilment to the commitment of 2030 Sustainable Development Agenda" (State Council of China 2016a). To achieve these aims and visions, the Healthy China 2030 Plan proposes five strategies: popularising a healthy life, optimising the health service, improving health protection, building a healthy environment, and developing a healthy industry. In actionable terms, the plan seeks to resonate with the WHO's Health in All Policies (HiAP) approach and reach high-income countries' primary health indicators by 2030.

5.2.5 A Multi-level, Nested Hierarchical System of Land-Use and Spatial Planning

This section introduces the Chinese system of land-use and spatial planning in three parts: (1) the five-level administrative system and processes for land-use and spatial planning, (2) the typology of land and spatial plans, and (3) the definition and standards of urban green spaces in China.

5.2.5.1 Five-Level Administrative System and Processes for Land-Use and Spatial Planning

This section introduces the Chinese land-use and spatial planning systems as they interact closely within the nested five levels of government. China's land-use and spatial planning system comprises five levels of administrative departments at the national, provincial, city, county, and township levels. After the national government's institutional reform in 2018, at the national level, the Ministry of Natural Resources and the Ministry of Housing and Urban–Rural Development coordinated land-use and spatial planning (see Fig. 5.8). At the provincial level, the Bureau of Natural Resources and Housing and Urban–Rural Development coordinates. The city level (municipality and autonomous region) is coordinated by the Bureau of Planning and Natural Resources and the Bureau of Housing and Urban–Rural Development.

Similarly, the district/county level is coordinated by the Division of Natural Resources and Planning and the Division of Housing and Urban–Rural Development. No offices are set expressly for managing natural resources and spatial planning at the township level; instead, the work is shared between the Economic Development Office and the Urban–Rural Development Office at this level. In this five-level administrative system, the superior-level agencies provide professional guidance and supervision to the lower levels. Land-use and spatial planning processes include planning, approval, implementation, supervision, and inspection/assessment. The process involves institutions of technical planning, administrative and specialist approval according to policies, and a set of technical standards, implementation, supervision, and regulations.

For forestry and landscaping, the Forestry and Grassland Bureau, subordinate to the Ministry of Natural Resources at the national level, is responsible for developing and conserving forest and grassland resources. Similarly, a Bureau of Forestry is established at the provincial level under the Bureau of Natural Resources and Planning to manage forestry production and conservation. At the city level, a Municipal Bureau of Forestry and Landscaping is set up, independently from the Bureau of Natural Resources and Planning, to manage forestry production, conservation, and Urban green spaces management. At the district level, forestry and landscaping-related work is directed by the Municipal Bureau of Forestry and Landscaping. At the township level, forestry-related work is handled by the Office of Economic Development, while township greening and landscaping-related work is dealt with by the Office of Urban and Rural Development instead. No departments are set up specifically for forestry and landscaping at the district and township levels.

Before 2018, the Ministry of Environmental Protection coordinated the work on environmental protection from central to local levels. Since the national government's institutional reform in 2018, the Ministry of Environmental Protection has been renamed the Ministry of Ecology and Environment, and the subordinate levels have also been renamed, assuming similar functions within their respective administrative areas. Similar to the work on urban greening and landscaping, at the township level, for the time being, no departments are set up specifically for environmental protection/ecological conservation. Instead, the work is shared between the Office of Economic Development and the Office of Urban and Rural Development.

5.2.5.2 Typology of Land and Spatial Plans in China

There are three types of land by the use status classified in the Chinese national standard *Current Land-Use Classification* (ref. GB/T21010-2017, in Chinese: 土地利用现状分类; tǔ dì lì yòng xiàn zhuàng fēn lèi): (1) farming lands, such as paddy fields and orchards; 2) development lands, such as commercial land, residential land, parks, and green spaces; 3) non-development lands, such as river surface, marshland, and glacier. This classification indicates that parks and green spaces are of a public service nature. Secondly, within the jurisdiction of a city, urban land is divided into development land and non-development land according to the current land classification standard *Urban Land Classification and Standards of Planning and Construction* (ref. GB50137-2017, in Chinese: 城市用地分类与规划建设用地标准; chéng shì yòng dì fēn lèi yǔ guī huà jiàn shè yòng dì biāo zhǔn). Urban development land includes residential, administration and public services, commercial and business facilities, industrial and manufacturing, logistics and warehouse, road, street and transport, municipal utilities, green space and square, and land to be defined. Urban non-development land includes water, farming, forestry, and other non-development areas. Again, Urban green spaces are considered as a type of urban development land in China (see Table 5.4).

There are several types of spatial plans in the Chinese planning system, including plans of primary functional areas, land-use plans, and urban and rural development plans. Urban and rural development plans then also include several types: city system plans (macro-level), master plans (city level), district plans (for big cities only), detailed plans (detailed development control plans), detailed construction plans, and special plans. Under the new government institutional framework, the Ministry of Housing and Urban–Rural Development, the Ministry of Natural Resources, and the National Development and Reform Commission prepare the national city system plans for planning at the subordinate levels. Provincial governments compile provincial urban system plans. City, district/county, and township governments prepare master plans for their administrative areas, respectively. In addition to master plans, governments at the city, county/district, and township levels also compile detailed

Table 5.4 Types of urban land and classification by function in China

Urban land	Urban development land	Urban non-development land
Classification by function	1. Residential, administration, and public services 2. Commercial and business facilities 3. Industrial and manufacturing 4. Logistics and warehouse 5. Road, street, and transport 6. Municipal utilities 7. Green spaces and squares including **urban parks, green buffers**, and public squares 8. Lands to be defined	1. Water area (excluding water bodies within urban green spaces and squires) 2. Farming and forestry area 3. Other area

development control plans per the requirements of the master plan within their respective administrative areas. The plans mentioned above are all statutories. Conceptual planning as a new type of urban plans has emerged in recent years as a more forward-looking plan incorporating new concepts which do not contain legal effects but serve as a reference of urban development and planning tendencies for planning professionals and policymakers.

5.2.5.3 Definition and Standards of Urban Green Spaces in China

As briefly mentioned above, the current Chinese land classification standard *Urban Land Classification and Standards of Planning and Construction* (ref. GB50137-2017) defines Urban green spaces as components of urban development land. Compared to an older version (ref. GBJ137-90) of the same land classification standard, there are changes in the terminology of Urban green spaces in relation to the population and in the minimal standards (see Table 5.5). Two things to note: first, there is an improvement in inclusiveness from measuring Urban green spaces only for the population residing in the city's central area (2011) to measuring it for the residents within the boundary of the development land of the city (2017). However, permanent residents in the Chinese policy documents refer to residents associated with an urban household registration. Instead, those associated with a rural household registration are considered temporary residents, which are not included in Urban green spaces size per capita measurements.

As for per capita green park areas, different city development schemes hold different standards. These schemes function on a voluntary base which cities apply depending on their development preferences. For example, for urban green area development, there are three schemes: *Garden City, Ecological Garden City,* and *Ecological City*, the standards of which increase in turn (see. In comparison, for example, the European Common Indicator of local public open areas does not set a target and is not explicitly focused on green space but is based on two criteria: (1) the

Table 5.5 Comparison of green space per capita requirements in earlier and current versions of *Urban Land Classification and Standards*

Urban Land Classification and Standards of Planning and Construction	
Version 2011, ref. GBJ137-90	Version 2017, ref. GB50137-2017
Green space per capita: the total green area in a city divided by the number of permanent residents in **the central area of the city**, in m^2 per capita	**Green space and public square area per capita:** the total green area and public squares in a city divided by the number of permanent residents **within the boundary of development land of the city**, in m^2 per capita
Park green area per capita: the total green area of parks in a city divided by the number of permanent residents in **the central area of the city**, in m^2 per capita	**Park green area per capita:** the total green area of parks in a city divided by the number of permanent residents **within the boundary of development land of the city**, in m^2 per capita
Minimum required: park green area per capita should be at least 8.0 m^2	Minimum required: park green area per capita should be at least 10.0 m^2

Table 5.6 Size-related standards for Urban green spaces planning in China and in Europe

Chinese cities	National Garden City Standards (in Chinese: 国家园林城市系列标准)	Park green area per capita no less than 9m^2 for cities with average development land per capita larger than 105 m^2 (alias the case of Guangzhou)
	National Ecological Garden City Standards (in Chinese: 国家生态园林城市标准)	Park green area per capita no less than 12 m^2 for cities with average development land per capita larger than 105 m^2
	National Urban Ecological Protection and Construction Plan (2015–2020) (in Chinese: 全国生态城市保护与建设规划2015–2020年)	Park green area per capita no less than 14.6m^2
	Urban development goals for Ecological Civilisation in 2020 and 2030	Park green area per capita no less than 17 m^2 and 20 m^2 respectively
European cities	Distribution: The percentage of citizens living within 300 m from a public open area of minimum size 0.5 ha Accessibility: approximate a 15-min walk	

percentage of citizens living within 300 m from a public open area of minimum size 0.5 hectares and (2) a distance that represents accessibility—approximate a 15-min walk.[4]

[4] Aambiente Italia Research Institute (2003).

5.3 Three Urban Green Spaces Governance Cases in Guangzhou

This section introduces the three Urban green spaces governance cases in Guangzhou. It includes an overview and the three respective cases, each containing the urban planning backgrounds, reasoning for intervention, the interventions, the results, the actors, the timelines, and their respective relevance as a case study.

Overall, Guangzhou has a basic Urban green spaces structure containing three main types of urban green space resources. The first type is a network of seven major ecological corridors distributed across the city. The second type is the major ecological hotspots, such as Yuexiu Park, Baiyun Park, and the Haizhu Wetland Park—the latter two are also referred to as the "Northern Lung" and the "Southern Lung" of the city, respectively. The third type includes the small green spaces scattered between other urban infrastructures. Having had a basic Urban green spaces structure in place, recent research points out many nuanced problems associated with ecosystem performance and spatial distribution. Small patches, heavy fragmentation, and random distribution characterised the vegetation landscape in Guangzhou's old urban areas. These problems could be an opportunity window if appropriate institutional arrangements are applied.

In contrast, the new urban areas were characterised by big patches and even distribution. Compared to natural forests, these Urban green spaces in Guangzhou also have lower species diversity and organic matter content (Guan and Chen 2003; Li, et al. 2000). For effective climate change adaptation, biodiversity, primary production, and ecological function of these urban green resources need to be increased.

5.3.1 Overview of the Three Case Studies

The three Urban green spaces governance cases in Guangzhou are: (1) Panyu Ecological Corridor (PEC), (2) Haizhu Wetland Park (HWP), and (3) Tianhe Sponge City demonstration site (TSCD) (see Fig. 5.11), which showcases the location of these three cases in Guangzhou City. These three cases occurred across the span of the last twenty years, each representing different institutional settings, namely, different combination of preferences, formal and informal, in urban development, property rights bundles, and governance arrangements. The basic characteristics of these three cases are presented in Table 5.7.

1. **Panyu Ecological Corridor (PEC)** case presents a conflict of interest between municipal and district governments. In this case, the Municipal Government of Guangzhou (GZ) aimed to promote a city-wide ecological corridor system, contradicting the economic development interest of the district government of Panyu (PY) which would require the land nearby or even within the PEC system.

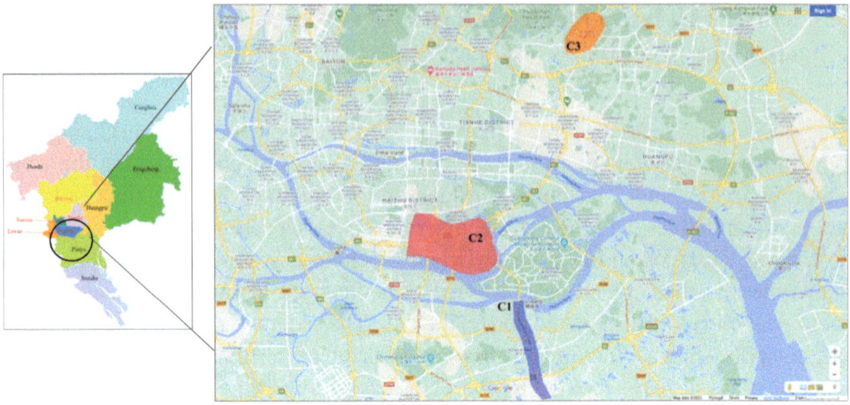

Fig. 5.11 Location of the three urban green space governance cases in Guangzhou, C1 = Case 1. Panyu Ecological Corridor, C2 = Case 2. Haizhu Wetland Park, C3 = Case 3. Tianhe Sponge City demo site. *Source* Based on Google Map

Table 5.7 Characteristics of the three Urban green spaces planning and governance cases in Guangzhou

Characteristics	Cases			
	C1: Panyu eco-corridor		C2: Haizhu Wetland Park	C3: Tianhe Sponge City demo site
Period	2000–2006		2012–2015	2015–2017
Size (km^2)	No data		3.67 km^2	3.57km^2
Administrative body		Guangzhou Municipal Government		
	Panyu District Government		Haizhu District Government	Tianhe District Government
Actors involved or influenced	Panyu District Government Business/individuals with interests in land Residents in neighbouring area		Haizhu District Government Villagers/Orchard land collective owners Urban migrant workers	Tianhe District Government Other departments in the municipal government Residents in neighbouring area
Method	Landscape adjustment/conversion		Land property right acquisition Landscape adjustment/ conversion	landscape adjustment/ conversion
Urban planning basis	Guangzhou Ecological conceptual plan		Guangzhou Master Plan	National Sponge City Initiative (under *Ecological Civilisation*)

2. **Haizhu Wetland Park (HWP)** case represents a conflict of interest between GZ, farmers (land- and homeowners), and urban migrants (tenants). The land property transfers only benefited the farmers but not the urban migrants due to the restriction imposed by the dual urban–rural household registration system, leaving urban migrants' quality of life poorly attended.
3. **Tianhe Sponge City demonstration site (TSCD)** case represents a conflict of interest between different branches within the municipal government, e.g., between the Bureau of Water Affairs and the Administration of Forestry and Gardening. In addition, for the private equipment and service suppliers with established connections to the conventional engineering solutions through public infrastructure construction projects, their vested interests are challenged by the Sponge City Initiative as it is a nature-based rather than engineering-based solution.

5.3.2 Panyu Ecological Corridor

The Urban Planning Background Due to the rapid expansion, land consumption in Chinese cities often outpaces the planned land-use indicators defined in master plans, invalidating them before the planning periods' end. In this context, the planning of urban green spaces is often in a relatively weak position that often gets compromised by the urgent land needs for other types of infrastructures for economic development. Such is the case for the Panyu District Ecological Corridor in Guangzhou. Panyu was a county, a sublevel unit of cities by administrative category, independent of Guangzhou City before the New Millennium. Considering the rapid urbanisation rate and land needs for future economic development, Guangzhou City merged with the former Panyu County in 2000, making Panyu a district of Guangzhou. Panyu District is in southern Guangzhou and the estuary of the Pearl River with a flat, alluvial landform ideal for urban sprawl. The district was, therefore, at risk of losing land resources with particularly critical ecological functions.

The Reasoning for Intervention From an ecologist's perspective, urban ecological corridors as essential components of urban green spaces not only because they perform the function of ecological evolution in nature but also play a supporting role for urban development by providing essential ecosystem services such as air, water and soil purification, noise cancellation, soil fixation, food production, education, recreation, and many more. Therefore, the principle of planning ecological corridors needs to respect and align with the spatial structure of urban construction areas to optimise their potential to provide ecosystem services.

The Intervention The Panyu Ecological Corridor, one of the seven major ecological corridors in Guangzhou (see Fig. 5.12), was laid out on a "conceptual plan" based on this principle. In 2000, Guangzhou's municipal government established a "strategic plan" for more sustainable urban development by planning, constructing, and conserving a city-wide ecological corridor network, which is the "Ecological

City" plan (Xinhua News Agency 2018). In the Chinese urban planning system, the terms "conceptual plan" and "strategic plan" refer to planning with a suggestive and non-legally binding nature, and they would be considered in more specific land-use plan-making. Conceptual and strategic plans are macro-scale, long-term plans in which future population growth projections and social and economic activities are considered. The conceptual plan of the ecological corridor network in Guangzhou (see Fig. 5.12) is composed of four horizontal (east–west) and three vertical (north–south) ecological corridors based on the major natural characteristics of the existing urban landscape. The three vertical (north–south) ecological corridors will go through the Panyu District. The one referred to as the Panyu Ecological Corridor in this research is the one planned alongside the S105 Nanshagang Expressway (S105 南沙港快速公路; Nán shā gǎng kuài sù gong lù).

The Result The third ecological corridor alongside the S105 Expressway (see Fig. 5.13) underwent constant erosion. This part of the ecological corridor was initially designed with a 500-m-wide green space on each side of the S105 Expressway. As planned without having any critical natural features basis such as mountains or waterways, but only based on a major concrete highway, the green spaces in this area were particularly exposed to the risk of erosion by the growing land demand for urban economic development. The fact that the conceptual plan was not legally binding meant the cost of contradicting other land interests and implementing the plan which was high. Consequently, various private construction proposals near the S105 Expressway were either formally or informally approved, gradually eroding the green space along the S105 Expressway. As a result, the initially designed 500-m-wide green space on each side was reduced to less than 100 m wide

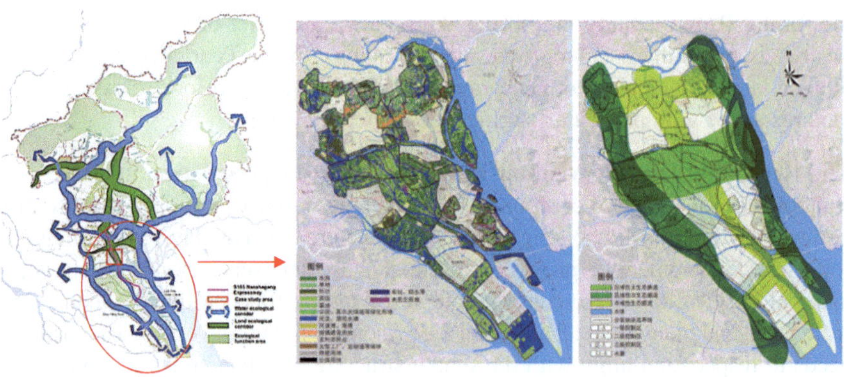

Fig. 5.12 Map of Panyu Ecological Corridor. *Left*: Structural map of the ecological corridor network in Guangzhou. *Middle*: Current status of land-use of Panyu District. *Right*: Planned ecological functional structure of Panyu District. *Source* 1. Master Plan, Planning and Construction Guideline of the Guangzhou Ecological Corridors, 2016 (in Chinese: 广州市生态廊道总体规划与生态廊道规划建设指引; Guǎngzhōu shì sheng tài láng dào zǒng tǐ guī huà yǔ sheng tài láng dào guī huà jiàn shè zhǐ yǐn). 2. Conceptual Plan of Panyu District Ecological Corridor (Qiao and Yang 2013)

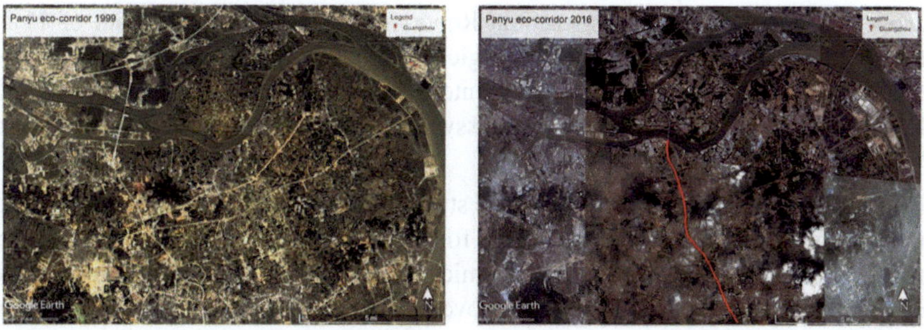

Fig. 5.13 Satellite image of the Panyu Ecological Corridor area prior and after planning, *Left*: prior to planning and construction of the S105 Expressway in 1999. *Right*: after planning and construction of the S105 Expressway in 2016. The S105 Expressway shown by the red line. © Google Earth

(see Fig. 5.13). The intended Panyu Ecological Corridor was eventually downgraded to the Nanshagang Expressway Green Belt, as referenced in recent policy documents ("南沙港快速路隔离绿带") (Guangzhou Forestry and Gardening Bureau 2012).

The Actors The actors and roles involved in this case study include:

(1) the Municipal Government of Guangzhou, which had the conflictual interests both to pursue economic progress as in macro, long-term development planning as well as to pursue the concept of "ecological city" by establishing an ecological corridor network throughout the city,
(2) the District Government of Panyu, which holds a primary interest in pursuing economic development opportunities,
(3) the private business initiatives which desire to acquire land nearby the ecological corridor for economic activities.

Since the conceptual plan had no legal rigidity, the low costs to disrespect the conceptual plan became a de facto incentive for the private sector or the citizens to erode the green spaces around the S105 Expressway and conduct economic activities.

The Timeline
2000: Panyu became a district of Guangzhou; the ecological corridor network conceptual plan of Guangzhou was proposed. A size standard proposed for ecological corridors is: 200-1000 m by width.

2009: A detailed development control plan[5] was made to reinforce the implementation of the parts of the ecological corridor network in Panyu District.

2015: A conceptual plan for eastern part of the ecological corridor network of Guangzhou was approved, specifying the controlled area of the S105 Expressway ecological corridor.

---

[5] Notice on the Controlled Detailed Plan of the Greening Corridors along the Four Expressways in Panyu District (in Chinese: 番禺区四条快速路绿化廊道控制性详细规划通告).

2018: An Implementation Assessment Report was indicating that 10 out of 11 in total ecological corridors in the ecological corridor network conceptual plan of Guangzhou were inadequately implemented, including the one alongside the S105 Expressway. The wording "S105 Expressway ecological corridor" was changed to "S105 Expressway isolation green belt."

Relevance as a Case Study In this case study, the ecological corridor plan's implementation had often been compromised for urban economic development practices for several reasons. First, placing economic development as the primary goal for the country has been consistent in China's governance history as a state since founded in 1949. The country aligns central-local efforts to economic development in a top-down administration system and a concentration of power. Second, Guangzhou's ecological corridor plan is conceptual, meaning that it does not have the legal rigidity of municipal master plans or regulatory plans, which are legally binding. In other words, as a conceptual plan, it was modified depending on the emerging needs or priorities in context. Therefore, the ecological corridor plan's implementation costs, namely, the designed 500-m-wide green space on each side of the S105 Expressway, were higher due to the often need for conflict resolution. Third, the ecological corridor plan's technical aspects—meaning the biodiversity and ecosystem services of the corridor—were not given sufficient attention or consideration. This was reflected by the lack of ecology experts in the planning and implementation processes, especially when conflicts of interest arose. Lack of political priority, legal formality, and technical expertise together constituted high costs for the Municipal Government of Guangzhou to complete the ecological corridor plan, especially for the one alongside the S105 Expressway.

5.3.3 *Haizhu Wetland Park*

The urban planning background: The Haizhu Wetland Park was initially an area of orchards referred to as the Guangzhou Wan Mu Orchard[6] run by farmers in the neighbour villages (see Fig. 5.14). The Guangzhou Wan Mu Orchard area is situated southeast Haizhu District, covering about 8 km2. As an urban wetland in the city's centre, the Wan Mu Orchard is essential to the city's ecological system. Therefore, the Orchard is also called the "Southern Lung" of Guangzhou, to the "Northern Lung"—the largest urban park in the city's northern area—the Baiyun Park. Traditionally, the Wan Mu Orchard had served as a major agricultural field for the neighbouring villages, which are now surrounded by the expanding urban construction and high-rise buildings in Guangzhou as the city is going through a rapid urbanisation process. These villages, therefore, are also referred to as urban villages.[7]

[6] The words Wan Mu literally means "ten thousand Mu," with Mu as a traditional Chinese standard unit for land area measurement: 1 Mu $\cong$ 666.67 m^2.

[7] In China, urban villages refer to a particular community form resulting from villages being surrounded rapidly by the expanding urban construction and high-rise buildings (Lan 2005; Liu

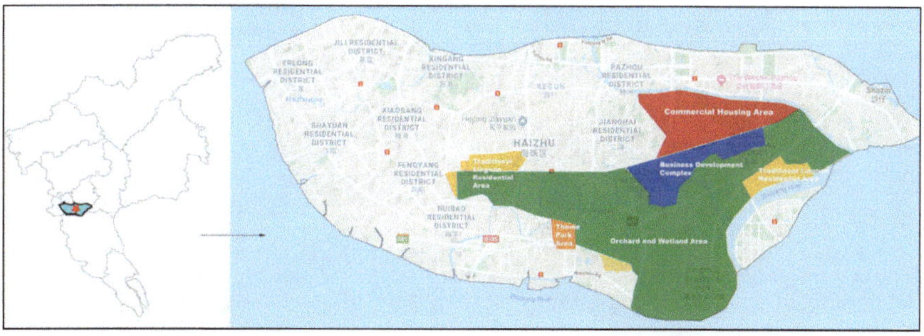

Fig. 5.14 Location and planning map of the Guangzhou Wan Mu Orchard Wetland Park

In the Haizhu Wetland Park case, the provision of benefits from establishing urban green spaces goes hand in hand with the living conditions of urban migrants. The land was transformed from farming land into a wetland park and acquired from nearby village farmers. Despite that financial compensation was made to villagers for their loss of agricultural income and that some land was preserved for farmers to explore future economic opportunities, appropriate planning, and infrastructure remain lacking in these villages as they remain collectively owned by villagers and are not converted into urban development land, which the local governments hesitate to develop. Also, informal housing continues to emerge in these villages, responding to a growing demand for affordable housing by urban migrants, who have overtaken the farmers and become the major inhabitants. Despite becoming de facto residents of Guangzhou, most of these migrants do not have access to public services such as free health care and education in Guangzhou which their urban peers have, since the rights to such services for these urban migrants are tied to the rural household registration of their places of origin. In this sense, the local governments have marginalised urban migrants residing in informal housing in the Haizhu wetland area.

The Reasoning for Intervention The above situation has generated several consequences. First, the Orchard has been eroded by urban infrastructures as the neighbouring land demand has been rising. The farmers made many temporary housing arrangements to meet the need of the growing number of urban migrants. Urban migrants gather in these urban villages as these housing options are more affordable. Although located within Guangzhou, the land of these urban villages and the Orchard is collectively owned by villagers and is not urban development land, which urban governments typically would have had the right and intention to develop. Second, the Orchard area has been polluted due to the growing industrial factories and urban migrants residing there without systematically planned grey infrastructure. Third,

---

and Gatzweiler, 2020). Although located within cities, the land of urban villages is collectively owned by villagers and is not urban development land. Therefore, the costs to build houses in urban villages are much lower than on what is legally defined as urban development land. For this reason, housing costs, in general, are much lower, forming a primary factor attracting urban migrants to inhabit those villages.

Fig. 5.15 Satellite image of the Haizhu Wetland Park area prior and after conversion *Left*: prior to conversion in 2009. *Right*: after land transfer and conversion in 2019. © Google Earth

the temporary housing arrangements provided greater benefits for the farmers than traditional agricultural practices, further accelerating the Orchard area's ecological deterioration.

The Intervention To address the environmental degradation problem and preserve the Orchard area as a key urban ecological hotspot, the Municipal Government of Guangzhou acquired the land ownership right of 90% of the Orchard from the villagers, changing it into state/public ownership, alias the new Haizhu Wetland Park, but keeping the farming land function to conserve the ecological potentials. The newly Haizhu Wetland Park is designed to be a semi-natural wetland-orchard hybrid ecosystem with a total size of 367 hectares (or 3.67 km^2, see Fig. 5.14). As a result of the 90% land property right transfer, the villagers/farmers were compensated financially with a total sum of 4.585 billion RMB (~ 6 million Euro, Municipal Government of Guangzhou, 2018) and the right to develop the remaining 10% collective land. The Wetland Park project has received 16 million RMB from the central government as subsidies for wetland protection and conservation. Till 2018, the Wetland Park project has costed the Municipal Government of Guangzhou and the District Government of Haizhu 6.3 billion RMB for the land property transfer, construction, and maintenance (Municipal Government of Guangzhou 2019a).

The Institutions Besides existing government institutions, new institutions were also created to improve the governance of Haizhu Wetland Park. The Haizhu District Government coordinated the Haizhu Wetland Park project. The land acquisition, planning, and construction were mainly conducted by the Haizhu District Housing and Construction Bureau and the Water Authority and collaborated with the Bureau of Landscape and Gardening. An administrative body named the Haizhu Wetland Management Office which was explicitly established to manage the Wetland Park. Besides, a research entity called the Haizhu Wetland Ecological Science Park was also co-established with a local university—the Sun Yat-Sen University, to foster scientific research.

The Timeframe

2009–2011: The Municipal Government of Guangzhou confirmed the ecological significance of the Haizhu Wan Mu Orchard for the city and decided to take actions to protect it. An international call was launched for the best proposals to redesign the Orchard. A winning proposal co-designed by the Guangzhou Urban Planning and Design Survey Research Institute and the Hangzhou Landscape Architecture Design Institute was selected.

2012: The State Council approved the property right transfer approach proposed by the Municipal Government of Guangzhou. An initial budget of 4.585 billion RMB was approved for the land acquisition and relocation of some urban migrants. The cost was shared between the Municipal Government of Guangzhou (80%) and the District Government of Haizhu (20%). A work plan for the land acquisition was formulated, and a bill was made to regulate the use of the budget—the land acquisition of 34 thousand people in 11 thousand households across the eight villages neighbouring the Orchard. The first phase of the Wetland Park was delivered.

2013: The Haizhu Wetland Management Office was established.

2014: The second phase of the Haizhu Wetland Park was completed.

2015: The Ministry of Forestry approved the park as a national-level wetland park.

2016: The third and fourth phases were completed. Meanwhile, the State Council formulated and promoted the national implementation of the *Proposal for Institutionalizing Wetland Protection and Restoration* (State Council of China 2016).

2018: The *Wetland Rehabilitation Plan for the Guangzhou Haizhu National Wetland Park* was compiled to formalise operational guidelines, standards, and targets for the Haizhu Wetland Park.

Relevance as a Case Study The central government evaluated Haizhu Wetland Park's conservation effort as successful in ecological conservation. The biodiversity at Wetland Park was recorded to have increased by 31.9% (District Government of Haizhu, 2016). The Haizhu Wetland Park transformation was also elected as one of the "40 innovations of local reforms in China's 40 years of economic reform and liberation" (Municipal Government of Guangzhou 2019b). There are several ways that Haizhu Wetland Park is relevant as a case study of institutional analysis of urban commons governance in China. First, the land acquisition strategy employed in this case directly deals with the issue of land property rights, and it reflects the city government's cogitation in reducing transaction costs. Second, by reserving the 10% collectively owned land for the villagers as their livelihood alternatives, the Municipal Government of Guangzhou could yet avoid another cost—the responsibility of rearranging their livelihood and the associated costs. Third, the urban migrants, who are the de facto majority of residents in these urban villages surrounding the Orchard, have been impacted negatively as the essential urban infrastructures and public services in these urban villages remain poorly developed, the available housing also became more expensive as their supply could no longer be increased, and density increased. As a result, the overall living and environmental quality of the urban migrants has deteriorated. These negative impacts on the urban migrants are factually hidden or neglected costs of the Wetland Park transformation, which even the newly created institutions of the Wetland Park could not address.

5.3.4 *Tianhe Sponge City Demonstration Site*

The Urban Planning Background In recent years, waterlogging, or urban inundation, has increasingly become a challenge in many Chinese cities due to rapid urban expansion and the layout of grey infrastructure. The current approach to urban inundation—using hydro-engineering means for fast discharge—has been found not only costly but ineffective for resolving the problem (Xia et al. 2017). The increasing deterioration in the urban environment and climate change impacts in Guangzhou determine that climate adaptation measures are urgently needed.

The Reasoning for Intervention While the subtropical monsoon climate characteristic brings heat- and water-related climate change impacts primarily, it also highlights the potential of a nature-based climate adaptation approach. Such a climate facilitates the rapid development of green infrastructure. Guangzhou's humid subtropical monsoon climate is an exogenous biophysical factor that facilitates the local governments in Guangzhou to adopt a nature-based approach for climate change adaptation (see Fig. 5.16).

The Intervention The central government took the first systematic action termed "the Sponge City Initiative" in 2013 to embrace urban green infrastructure as a Nature-based Solution to climate change impacts (see Table 5.8). This initiative

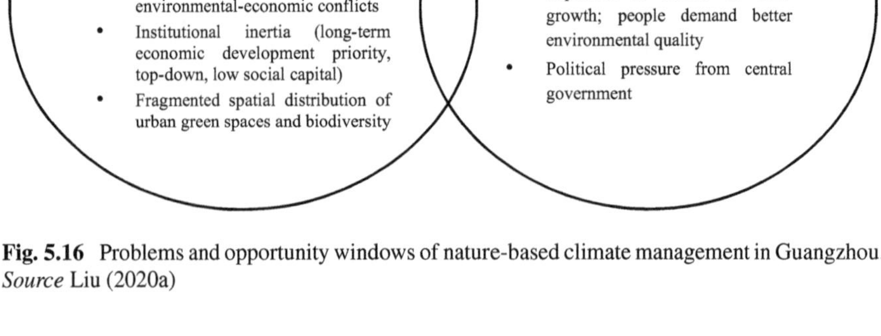

Fig. 5.16 Problems and opportunity windows of nature-based climate management in Guangzhou. *Source* Liu (2020a)

refers to actions that adopt new urban green infrastructure or improve existing ones to enhance ecological functions, particularly stormwater absorption, storage, retention, purification, and release for surface runoff control (Liu and Sun 2017). The Sponge City Initiative endorses three approaches: (1) conserving existing major urban parks and green spaces, which deliver essential ecosystem services; (2) restoring degraded ecosystems; and (3) low-impact development (LID). While the first two approaches are ecosystem-based, the third is characterised primarily by low-impact engineering-based interventions. The basic standard for becoming a Sponge City is to absorb, retain, store, and utilise 70% of the stormwater within an urban territory by applying the three intervention means mentioned above. The initiative sets the goals that cities should have at least 20% and 80 of urban built-up areas that meet the standard by 2020 and 2030. At the central level, the implementation is led by the Ministry of Housing and Urban–Rural Development (MOHURD), with financial support from the Ministry of Finance and technical support from the Ministry of Water Resources.

Notably, even though the central government has gradually made the Sponge City Initiative an imperative for big cities and has chosen two groups of pilot cities in 2015 and 2016, Guangzhou was not in the two groups of pilot cities. On a voluntary basis, Guangzhou also has adopted the Sponge City Initiative and established the *Special Plan of Sponge City Initiative Implementation in Guangzhou 2016–2030* (*Special Plan* for short, see Fig. 5.17) under an extensive legal and policy framework. The framework includes the National Environmental Protection Law, the National Water Law, the National Urban and Rural Planning Law, the State

Table 5.8 Overview of the Chinese Sponge City Initiative

Main attributes	Characteristics
Launch	2013
Responsible government branches	Lead: Ministry of Housing and Urban–Rural Development (MOHURD) Financial support: Ministry of Finance (MF) Technical support: Ministry of Water Resources (MWR)
Approaches	Conserving existing major urban parks, green spaces Restoring degraded ecosystems LID
Standard	Being able to absorb, retain, store, and utilise 70% of the stormwater within its urban territory Indicator: annual rainfall-runoff rate control
Goals	By 2020, 20% of urban built-up areas meets the standard By 2030, 80% of urban built-up areas meets the standard
Pilot cities	First group: 16 cities in 2015 Second group: 14 cities in 2016
Finance	Funding from the central government for each pilot city varies from 400 to 600 million RMB (57–85 million US$) per year Implementation via public–private partnerships (PPPs)

Council Guideline on Sponge City Construction, among others. Implementing the Sponge City Initiative on a voluntary basis means that Guangzhou must finance its Sponge City Construction independently. Thus, local officials were encouraged to explore financing possibilities for implementation.

The Institutions At the local level, the Sponge City Initiative in Guangzhou is implemented by the Leading Group of Sponge City Construction, primarily led by the municipal government and participated by the Bureau of Housing and Urban–Rural Development, the Bureau of Finance, and the Water Agency. Each district is then equipped with the same organisational structure to prioritises implementation. The municipal government aimed to implement the *Special Plan* within a focus urban area of 1092.5 km2. The district governments are then responsible for implementing it within their respective territory. The *Special Plan* also sets various technical guidelines and standards for quality inspection, ranging from urban water supply standards to outdoor drainage, urban greening, vertical greening, and flood control standards. These standards constitute the information necessary for implementing the Sponge City Initiative effectively. The *Special Plan* makes local officials accountable for the implementation by making it a key performance indicator (KPI) with possible sanctions in case of non-compliance, including non-reward or demotion.

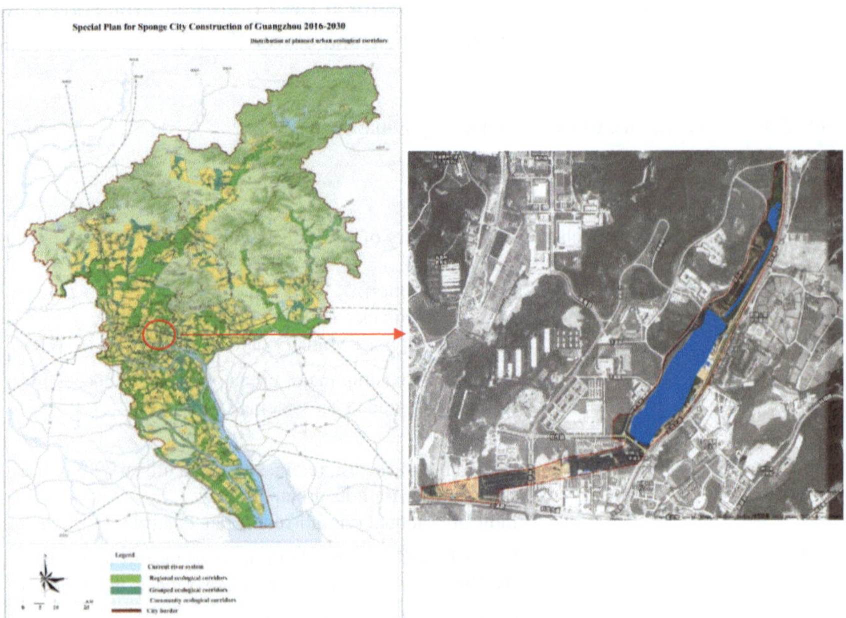

Fig. 5.17 Guangzhou Tianhe Sponge City demonstration site. *Left*: Special Plan for Sponge City Construction of Guangzhou 2016–2030. *Right*: Intervention area of the Tianhe Sponge City demo site. *Source* Guangzhou Urban Planning & Design Survey Research Institute. *Source* www.turenscape.com

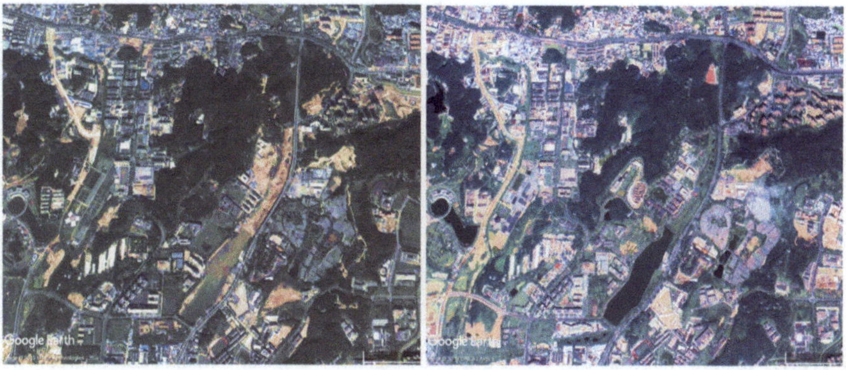

Fig. 5.18 Tianhe Sponge City demonstration site satellite images, prior and after conversion, *Left*: prior to conversion in 2013. *Right*: after conversion in 2019. © Google Earth

The Sponge City Demonstration Site discussed in this research involves a wetland park that was converted from abandoned water pools and sweet water aquaculture ponds next to the newly developed Tianhe Smart Industrial Park (in Chinese 天河智慧城, see Fig. 5.19). The Leading Groups of Sponge City Construction commissioned a private urban landscape design consulting company to transform this area based on landscape design techniques that are adherent to the national Sponge City policy guidelines. The demonstration site is an important part of the natural water corridor in Tianhe District, covering an area of 46.8 hectares (see Fig. 5.17). The low-lying site comprises a reservoir, farmland, and aquaculture ponds. The demonstration site is aimed to address four challenges:

1. to deal with the local natural climate (large annual rainfall and frequent torrential rains highly prone to waterlogging),
2. to reduce water pollution in the Tianhe area and improve the water quality,
3. to create a water-adaptable plant community for adaptation to the drastic drought and flood alternations,
4. to provide an open space that meets the leisure needs of surrounding residents.

The Timeframe

2012: The "Sponge City" concept was first proposed at the Low Carbon Cities and Regional Development Science and Technology Forum.

2013: The Ministry of Housing and Urban–Rural Development (MOHURD) launched the Sponge City Initiative at the national level.

2014: MOHURD launched the *Sponge City Construction Technical Guide.*

2015: The first group of pilot cities, 16 in total, was confirmed and given financial support for implementing the Sponge City Initiative.

2016: The second group of pilot cities, 14 in total, was confirmed and given financial support for implementing the Sponge City Initiative. Guangzhou launched the

Fig. 5.19 Images of Sponge City Demo Site in Guangzhou—Daguan Wetland Park. *Source* www.turenscape.com

Special Plan of Sponge City Initiative Implementation in Guangzhou 2016–2030 on a voluntary basis.

2016: The first Sponge City demonstration site—Daguan Wetland Park in Tianhe District, was concluded.

2017: The central government proposes that all cities should take the initiative to build Sponge Cities in the Annual Work Report.

2020: The Guangzhou Municipal Government sets up an expert committee for supervising Sponge City Construction.

Relevance as a Case Study The Sponge City Demonstration Site (TSCD) case represents a conflict of interest between different branches within the Municipal Government of Guangzhou, e.g., between the Bureau of Water Affairs and the Administration of Forestry and Gardening. In addition, the equipment and service suppliers in the private sector with established connections to conventional engineering solutions through public infrastructure construction projects and their vested interests are challenged by the Sponge City Initiative as it is a nature-based rather than an engineering-based solution. This is thus an urban green space governance case which showcases the interaction dynamics of multiple actors, within the public and between public and private ones, and how institutions at the national and local levels evolve.

References

Aambiente Italia Research Institute (2003) European common indicators: towards a local sustainability profile. Final report. Milano, Italy. Available at http://www.cityindicators.org/Deliverables/eci_final_report_12-4-2007-1024955.pdf. Accessed on 19 May 2016

ACCC (2012) Managing water as China warms: new insights from regional models. Key research findings. Available at: https://www.weadapt.org/sites/weadapt.org/files/2017/january/accc_1_water.pdf

Asian Development Bank (2015) Addressing climate change risks, disasters, and adaptation in the People's Republic of China. ISBN:978-92-9257-156-6
Balaguru K, Foltz GR, Leung LR, Emanuel KA (2016) Global warming-induced upper-ocean freshening and the intensification of super typhoons. Nat Commun. https://doi.org/10.1038/ncomms13670
Berry HL, Bowen K, Kjellstrom T (2010) Climate change and mental health: a causal pathways framework. Int J Public Health 55(2):123–132. https://doi.org/10.1007/s00038-009-0112-0
Carson R (2000) Silent spring (Reprint). Penguin UK
Chen M, Liu W, Lu D (2016) Challenges and the way forward in China's new-type urbanization. Land Use Policy 55:334–339. https://doi.org/10.1016/J.LANDUSEPOL.2015.07.025
Cheng L, Chen X (2018) The development and innovation of the socialist market economy theory with Chinese characteristics in the past 40 years of reform and opening up (in Chinese: 改革开放40年中国特色社会主义市场经济理论的发展与创新). Econ Perspect (in Chinese: 经济学动态) 12. http://iar.sufe.edu.cn/_upload/article/files/18/8b/ef3d99ee4b0995a7f0a9cd7e4ab2/650105e7-f5f3-423c-b80f-89ec20660429.pdf
Cheng S, Xie G, Cao S, Kitzes J, Buchan S, Goldfinger S (2012. Report on China's ecological footprint (in Chinese: 中国生态足迹报告). https://www.footprintnetwork.org/content/images/uploads/China_Report_zh.pdf
Chinese National Bureau of Technical Supervision (1994) Standard for climatic regionalization for architecture (in Chinese: 建筑气候区划标准GB50178-93)
C40 Cities Climate Leadership Group (n.d.) Climate change adaptation in delta cities (in Chinese: 三角洲城市的气候变化适应指南), p 22. C40 Cities Climate Leadership Group
Dai F, Duan X (2003) Before and after an urban master development strategic plan—proposal and reflection on the strategic plan of Guangzhou (in Chinese: 城市总体发展战略规划的前前后后——关于广州战略规划的提出与思考). Urban Plan (in Chinese: 城市规划) 1:24–27
Dallmayr F, Zhao T, Ji W, Zhou L, Zhang F, Liu S, Chen M, He B, Ni P, Ci J, Cui Z, Wang S, Cheng G (2012) Contemporary Chinese political thought: DEcosystem-based adaptationtes and perspectives. University Press of Kentucky. https://www.jstor.org/stable/j.ctt2jcsvm
Deng X, Ran Y, Zeng K (2017) The evolution and reflection on the Urban Master Planning of Guangzhou (in Chinese: 广州城市总体规划的演变与思考). Urban Plan Forum (in Chinese: 城市规划学刊). Available at: https://upforum.tongji.edu.cn/f1/99/c11653a127385/page.htm
Elsne, JB, Kossin JP, Jagger TH (2008) The increasing intensity of the strongest tropical cyclones. Nature 455(7209):92–95. https://doi.org/10.1038/nature07234
Feng X, Tsimplis MN (2014) Sea level extremes at the coasts of China. J Geophys Res: Oceans. https://doi.org/10.1002/2013JC009607
Fritze JC, Blashki GA, Burke S, Wiseman J (2008) Hope, despair and transformation: climate change and the promotion of mental health and wellbeing. Int J Ment Heal Syst. https://doi.org/10.1186/1752-4458-2-13
Gong J, Hu Z, Chen W, Liu Y, Wang J (2018). Urban expansion dynamics and modes in metropolitan Guangzhou, China. Land Use Policy 72(October 2016):100–109. https://doi.org/10.1016/j.landusepol.2017.12.025
Guan D, Chen Y (2003) Status of urban vegetation in Guangzhou City. J Forest Res 14(3):249–252. https://doi.org/10.1007/bf02856840
Habitat UN (2016) World cities report 2016—Urbanization and development: emerging futures. Int J. https://doi.org/10.1016/S0264-2751(03)00010-6
Hallegatte S, Green C, Nicholls RJ, Corfee-Morlot J (2013) Future flood losses in major coastal cities. Nat Clim Chang. https://doi.org/10.1038/nclimate1979
Hu A, Hu L, Chang Z (2003) China's economic growth and poverty reduction (1978–2002). https://www.imf.org/external/np/apd/seminars/2003/newdelhi/angang.pdf
IPCC. (2014). Summary for policymakers. Climate change 2014: Impacts, adaptation and vulnerability—Contributions of the working group II to the fifth assessment report, pp 1–32. https://doi.org/10.1016/j.renene.2009.11.012

Lan Y (2005) Villages in cities: a field study of a “New Village Community”, 1st edn. Sanlian Lan Shanghai: SDX Joint Publishing Company.

Li Z Wang L, Guan D (2000) Landscape heterogeneity of urban vegetation in Guangzhou (in Chinese: 广州城市绿地系统景观异质性分析). J Appl Ecol (in Chinese: 应用生态学报), 11(1):127–130

Li D (2018) Three major theoretical issues in the construction of contemporary Chinese political philosophy (in Chinese: 当代中国政治哲学建构的三个重大理论问题). Soc Sci Digest (in Chinese: 社会科学文摘) 3. http://www.aisixiang.com/data/115062.html

Li Z (ed) (2016) Research on the theory, methods and practice of public policy evaluation, 2nd edn. China Development Press

Literature Research Office of the Central Committee of the Communist Party of China (2005) Selected documents since the 16th National Congress of the Communist Party of China, vol 1 (in Chinese: 十六大以来重要文献选编上). Party Literature Research Center

Liu T, Xu YJ, Zhang YH, Yan QH, Song XL, Xie HY, Luo Y, Rutherford S, Chu C, Lin HL, Ma WJ (2013) Associations between risk perception, spontaneous adaptation behavior to heat waves and heatstroke in Guangdong province. BMC Public Health, China. https://doi.org/10.1186/1471-2458-13-913

Liu J, Gatzweiler FW (2020) The institutional challenge to co-deliver migrant integration and urban greening—evidence from Haizhu Wetland Park Project in Guangzhou, China. J Chin Govern. https://doi.org/10.1080/23812346.2020.1760069

Liu M, Sun W (2017) Report on China’s Sponge City initiative (in Chinese: 中国海绵城市建设报告). In: The report on the development of China’s eco-cities (in Chinese: 中国生态城市建设发展报告), pp 360–372. Social Science Academic Press, China

Liu J (2020) Narrating the impacts of climate change for urban health governance in Guangzhou, China. Cities & Health. https://doi.org/10.1080/23748834.2019.1701945

Murphy C (2017). Hu line: China’s forgotten frontier. Sixth Tone. https://www.sixthtone.com/news/1000459/hu-line-chinas-forgotten-frontier

Marinelli M (2018) How to build a ‘Beautiful China’ in the anthropocene. the political discourse and the intellectual DEcosystem-based adaptationte on ecological civilization. J Chin Polit Sci. https://doi.org/10.1007/s11366-018-9538-7

Muis S, Verlaan M, Winsemius HC, Aerts JCJH, Ward PJ (2016) A global reanalysis of storm surges and extreme sea levels. Nat Commun. https://doi.org/10.1038/ncomms11969

Nakamura R, Shibayama T, EstEcosystem-based Adaptationn M, Iwamoto T (2016) Future typhoon and storm surges under different global warming scenarios: case study of typhoon Haiyan (2013). Nat Hazards 82(3):1645–1681. https://doi.org/10.1007/s11069-016-2259-3

National Climate Center (2018) Non-Party stakeholder submission to the Talanoa dialogue of the UNFCCC. https://unfccc.int/sites/default/files/resource/162_Talanoa_Dialogue_Submission_NCC_China_2018.pdf

Nicholls RJ, Hanson S, Herweijer C, Patmore N, Hallegatte S, Corfee-Morlot J, Château J, Muir-Wood R (2007) Ranking port cities with high exposure and vulnerability to climate extremes: exposure estimates. In: OECD environment working papers, no 1. https://doi.org/10.1787/011766488208

OECD (2006) Environmental compliance and enforcement in China—An assessment of current practices and ways forward. Available at: https://www.oecd.org/env/outreach/37867511.pdf

Pan J (2015) China’s environmental governing and ecological civilization. Springer, Berlin, Heidelberg. https://doi.org/10.1007/978-3-662-47429-7

People’s Daily (2008) The scientific connotation of people-centered development. The Scientific Outlook on Development. http://theory.people.com.cn/GB/40557/130316/130317/7600245.html

Wei L (ed) (2018) Research report on China’s institutional reform for ecological civilisation, 1st edn. China Development Press

Russell B (2009) The prospects of industrial civisation. Unwin Brothers Limited, Great Britain

Silva RA, West JJ, Lamarque JF, Shindell DT, Collins WJ, Faluvegi G, Folberth GA, Horowitz LW, Nagashima T, Naik V, Rumbold ST, Sudo K, Takemura T, Bergmann D, Cameron-Smith P, Doherty RM, Josse B, MacKenzie IA, Stevenson DS, Zeng G (2017) Future global mortality from changes in air pollution attributable to climate change. Nat Clim Chang 7(9):647–651. https://doi.org/10.1038/nclimate3354

State Council of China (1994) China's agenda 21: white paper on china's population, environment and development in the 21st century. China population today. China's 10th five-year plan for national socio-economic development (2001) (testimony of State Council of China)

Sun Q, Wang W, Wang Y, Li S (2018) Estimating and projecting the acute effect of cold spells on excess mortality under climate change in Guangzhou. Chin J Prevent Med (in Chinese: 中华预防医学杂志) 52(4):430–435. https://doi.org/10.3760/cma.j.issn.0253-9624.2018.04.018

Tracy A, Trumbull K, Loh C (2007) The impact of climate change in Hong Kong and the Pearl River Delta. China Perspect. https://doi.org/10.4000/chinaperspectives.1173

Walsh J, Wuebbles D, Hayhoe K (2014) Chapter 2: our changing climate. Climate change impacts in the united states: the third national climate assessment. October, pp 19–67. https://doi.org/10.7930/J0KW5CXT.On

Xia J, Zhang YY, Xiong LH, He S, Wang LF, Yu ZB (2017) Opportunities and challenges of the Sponge City construction related to urban water issues in China. Sci China Earth Sci. https://doi.org/10.1007/s11430-016-0111-8

Xiang J, Hansen A, Liu Q, Liu X, Tong MX, Sun Y, Cameron S, Hanson-Easey S, Han GS, Williams C, Weinstein P, Bi P (2017) Association between dengue fever incidence and meteorological factors in Guangzhou, China, 2005–2014. Environ Res. https://doi.org/10.1016/j.envres.2016.11.009

Xiao Y, Song Y, Wu X (2018) How far has China's urbanization gone? Sustainability (switzerland). https://doi.org/10.3390/su10082953

Xu C (2009) The position of Ecological Civisation in the human civisation. In: China Society for Hominology (ed) Ecological civisation, globasation and human development conference proceedings, pp 158–161

Yang HF, Yang SL, Xu KH, Wu H, Shi BW, Zhu Q, Zhang WX, Yang Z (2017) Erosion potential of the Yangtze Delta under sediment starvation and climate change. Sci Rep. https://doi.org/10.1038/s41598-017-10958-y

Zhou Q (2010) Research on construction land expansion based on GIS and RS in Guangzhou for nearly 20 years. Guangzhou University

Zhu Da, Ru J (2008) Strategic environmental assessment in China: Motivations, politics, and effectiveness. J Environ Manage. https://doi.org/10.1016/j.jenvman.2007.03.040

Zuo K (2014) Critical issues in the next decade of China's infrastructure effort. McKinsey & Company. https://www.mckinsey.com/featured-insights/urbanization/critical-issues-in-the-next-decade-of-china-infrastructure-effort

Chapter 6
Analysis

The previous chapter contextualised the biophysical conditions of the three case studies, outlined the institutional context, and introduced the three case studies, including the actors and a brief portrait of the local communities (particularly in the second case), which correspond to the first three exogenous and independent variables: ① *biophysical conditions*, ② *community attributes*, and ③ *rules-in-use* of the IAD framework (see Sect. 3.5 Analysing Common-Pool Resource Governance through the IAD Framework).

6.1 Characteristics and Tendencies of the Rules-in-Use in Case Studies

6.1.1 Characteristics of the Rules-in-Use in Case Studies

The analysis in this section explains how these policy documents were used as different rules to influence the outcome of the three Urban green spaces cases. The aim is to understand the institutional dynamics, facilitators, and underlying barriers Chinese local governments which have faced regarding the need and mandates to plan and manage Urban green spaces that could lead to more effective Nature-based Solutions implementation.

In Sect. 3.5 Analysing Common-Pool Resource Governance through the IAD Framework, *rules-in-use* are introduced as the information that forms a common understanding of those involved to who must, must not, or may take which actions affecting others subject to sanctions (Crawford and Ostrom 1995). The *rules-in-use,* alias, *institutions* studied in this research, refer to the formal plans, policies, and environmental agendas made by or agreed upon by the Chinese national government, the Municipal Government of Guangzhou, and planning institutes, as well as the

© Zhejiang University Press 2023

J. Liu, *Transforming Urban Green Space Governance in China Under Ecological Civilization: An Institutional Analysis*, Urban Health and Wellbeing,
https://doi.org/10.1007/978-981-99-6694-3_6

informal institutional factors such as central-local political dynamics, urban planning culture, and historical land tenure. The official policy documents and government reports both at the national and local levels that meet these two criteria are selected for analysis: (1) have cast direct or indirect impacts on the shaping, planning, and implementation of the three Urban green spaces cases in Guangzhou and (2) have documented these processes (for the complete list of these official policy documents and government reports, please refer to Annex 5. List of Policy Documents).

There are six main types of policy documents (see Annex 5. List of Policy Documents) in this analysis:

(1) **strategic plans**, which systematically set ambitions for the future and determine how best to achieve them,
(2) **action plans**, which help actors know what needs to be done to complete a task, project, initiative, or strategy related to Urban green spaces planning and implementation, including steps, milestones, and measures of progress, responsibilities, specific assignments, and timeline,
(3) **conceptual plans**, which are used to introduce novel urban planning concepts and communicate essential information on how to operationalise them. They signal a commitment to a diligent planning effort but are not legally binding,
(4) **regulation/technical guidelines**, which guide the technical operation of a particular urban plan, including material and human resource input and basic standards,
5) **government reports**, which are generated to communicate information that has been compiled as outcome of research and analysis of data and issues,
(6) **communique/official bulletins** are similar to government reports as they are periodical publications that are authorised to publish legal notices or announce achievements made around Urban green spaces planning and governance.

When distinguished by policy document type, the policy documents at the national and local levels as the rules-in-use in this research have the following characteristics:

- The strategic plans and action plans at both the national and local levels serve primarily as *scope rules* and *information rules* as they define the range of possible results by setting anticipated and binding targets and more general directions of development.
- Besides, local strategic plans and action plans also serve as *boundary rules* and *aggregation rules* as they define the geographic and demographic boundaries of actors and actions, as well as how actors interact (collaborate, obey, cooperate, monitor, and more) to achieve the action plans.
- Conceptual plans introduce novel urban planning concepts and communicate essential information on how to operationalise them. Therefore, they serve primarily as *information rules.*
- Regulation/technical guidelines function as *position rules*, *choice rules*, and *payoff rules*, as they define not only material resources but also human resources (actors) and their positions, choices, and possible sanctions if actors did not meet certain standards.

- Finally, reports and communique/official bulletins serve as summaries or feedback to the *scope rules* and the *payoff rules* as they provide data to show the results of a particular Urban green spaces intervention.

When distinguished by levels, the policy documents at the national and local levels differ in their focus. For example, when serving as *payoff rules*, the MaxQDA analysis shows the national-level policy documents as *payoff rules* primarily define formal institutional change (see Fig. 6.1). In contrast, local policy documents mostly describe known costs and benefits in social, environmental, and economic dimensions (see Fig. 6.2). This analysis considers codes related to costs and benefits and formal institutional changes related to *payoff rules*, which define rewards and sanctions.

A horizontal inspection on the impact distribution of these policy documents of the three Urban green spaces cases shows that, **the quantities of different types of policy documents vary in the three cases; in comparison, the second case has the most policy documents as support** (see Table 6.1). In the first case—the Panyu Ecological Corridor construction—there are five strategic plans and two action plans at the national level, two strategic plans and four at the local level, plus a conceptual plan but no regulation or technical guide. In the second case—the Haizhu Wetland Park project, there are as many as 11 strategic plans and eight action plans at the national level, five strategic plans and five action plans at the local level, plus a conceptual plan, and at the same time, there are five regulation/technical guidelines.

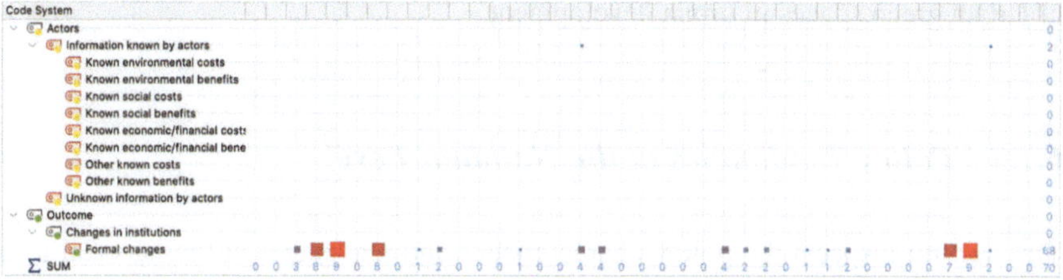

Fig. 6.1 Codes* on *known costs and benefits* and *formal institutional changes* and their distribution frequency in national-level policy documents. * Each coloured square at the nodes represents a certain number of coded segments. The larger the square, the more segments were assigned with the respective codes

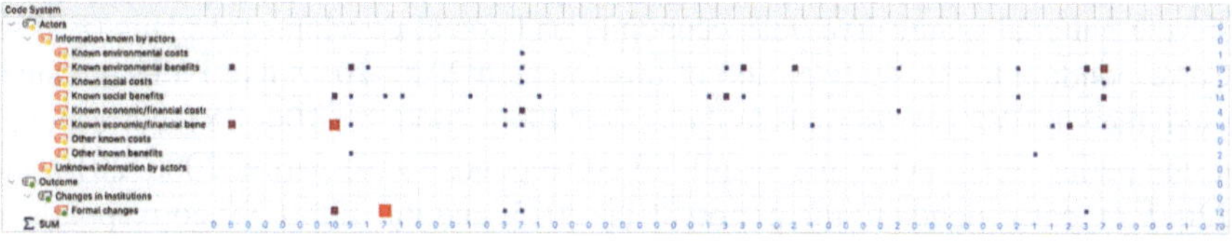

Fig. 6.2 Codes* on *known costs and benefits* and *formal institutional changes* and their distribution frequency in local-level policy documents. * Each coloured square at the nodes represents a certain number of coded segments. The larger the square, the more segments were assigned with the respective codes

Table 6.1 Ratio of national *versus* local policy documents in numbers in each type within the list of policy documents for each case study

Policy document type		Strategic plan	Action plan	Conceptual plan	Regulation/ tech. guideline	Report	Communique/ official bulletin
Cases	Case 1	5:2	2:4	0:1	0:0	1:15	2:0
	Case 2	11:5	8:5	1:0	4:1	2:18	1:0
	Case 3	5:4	5:5	1:0	6:4	4:15	1:0

In the third case—the Sponge City demonstration site—there are five strategic plans and five action plans at the national level, four strategic plans and five at the local level, and a conceptual plan. There are as many as ten regulations/technical guidelines (six at the national level and four at the local level). Many government reports, especially local ones, have concluded on developing the three Urban green spaces cases.

A vertical comparison in Table 6.1 shows that **the number of technical guidelines has increased from the first Urban green spaces case to the third Urban green spaces case, indicating that the state and local governments are probably paying greater attention to the technicality and legality of the operation of Urban green spaces planning and governance.** Besides, the second Urban green spaces case is associated with far more strategic plans and action plans than the first one, indicating that the national and local governments are paying greater attention to the importance of Urban green spaces in urban development.

6.1.2 Policy Tendencies at the National Level

This section inspects the policy characteristics and tendencies at the national level through four analyses: (1) the frequency of key policy terms, (2) articulation with international sustainable development agendas, (3) the intensity of key policy targets, and (4) shifting patterns of contemporary political philosophy. The aim is to understand the institutional dynamics, which might reveal the facilitators and underlying barriers to urban green space planning and governance in Guangzhou.

At the national level, the most widely overarching and recurring policy documents are the national macro five-year economic and social development plans (FYPs). Corresponding to the study period of this research 2000–2019, five FYPs are analysed: the 9th FYP corresponds to the period 1996–2000, the 10th FYP corresponds to the period 2001–2005, the 11th FYP corresponds to the period 2006–2010, the 12th FYP corresponds to the period 2011–2015, and the 13th FYP corresponds to the period 2016–2020. Figure 6.3 shows the changes in the frequencies of the four key terms "Low carbon (低碳 dī tàn)," "Greening (绿色 lǜ sè)," "Climate change (气候变化 qì hòu biàn huà)," and "Sustainable Development (可持续发展 kě chí xù fā zhǎn)" mentioned the five FYPs. Figure 6.3 shows a steady increase in the

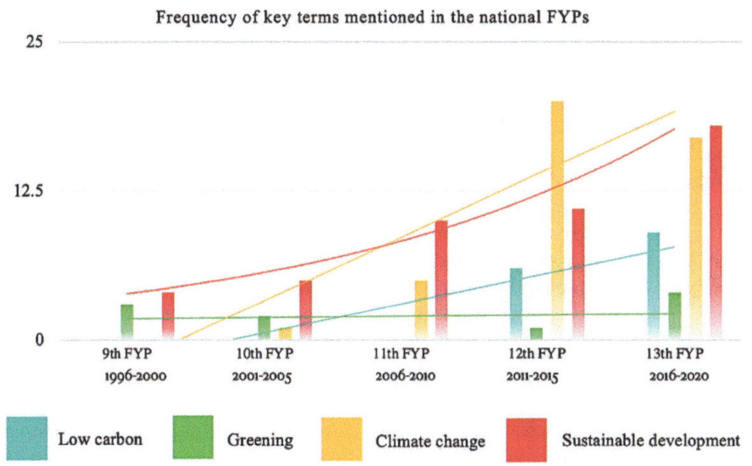

Fig. 6.3 Frequency of key terms mentioned in the national FYPs

frequencies of the term "Sustainable development" and "Low carbon" and a sharp increase in "Climate change" in the national FYPs.

Another critical aspect of Urban green spaces policy tendency at the national level lies in how actions are defined at this level based on international sustainable development agendas. This research finds **several direct interventions on or related to Urban green spaces that the Chinese national government plans to take to fulfil the 2030 Agenda for Sustainable Development**. Six targets in three SDGs: Goal 11—Sustainable Cities and Communities, Goal 13—Climate Action, and Goal 15—Life on Land are responded to with action plans by the Chinese national government (see Table 6.2).

In addition, the intensity of key policy targets, including anticipated and legally binding targets, within the most recent three FYPS is inspected. Anticipated targets are development goals that the government anticipates without legal or political obligation to achieve. Binding targets are development goals based on results expected that further clarify and strengthen government responsibilities and are required to be accomplished by the local governments. The analysis, showcased in Fig. 6.4, suggests that **from the 11th FYP to the 13th FYP, there are more legally binding targets (green) and less anticipated targets (blue) on resource and environment, as well as less anticipated targets on economic growth, indicating a more robust political will of the Chinese central government for environmental governance and shifting out of the economic growth focus**. The number and ratio between anticipated and binding Public Service and Well-being targets are about the same across the three FYPs. The analysis also showcases an increasing focus on Technology and Innovation.

Furthermore, how contemporary political philosophy is narrated in national policy plans was also examined. Figures 6.5, 6.6, and 6.7 showcase the distribution frequency of contemporary political philosophy in the five national FYPs, National Climate Change and Sustainable Development Plans, and National Urban Spatial Planning and Development Plans, respectively. These three figures indicate **a similar**

Table 6.2 Direct interventions on or related to urban green spaces explicated in China's *National Plan on Implementation of the 2030 Agenda for Sustainable Development*

SDGs and targets		Actions taken in China
Goal 11. Sustainable Cities and Communities	**11.6** By 2030, reduce the adverse per capita environmental impact of cities, including by paying special attention to air quality and municipal and other waste management	**Promote urban and rural green spaces actively and increase the size of green space per capita.** Improve the management of urban domestic waste, promote rural domestic waste treatment, and improve waste treatment quality. Develop a plan for achieving urban air quality standards. By 2020, the number of days of heavy pollution in prefecture-level and above cities should be reduced by 25%
	11.7 By 2030, provide universal access to safe, inclusive, and accessible, green, and public spaces, in particular for women and children, older persons, and persons with disabilities	Control urban development intensity strictly and **protect urban and rural green and ecological spaces. Promote urban green belts and ecological corridors** combined with the **restoration of water bodies and wetlands**, road traffic systems, and the protection of scenic resources. By 2020, the green space rate per urban built-up area should reach 38.9%, and green space per capita should reach 14.6 m^2
Goal 13. Climate Action	13.1 Strengthen resilience and adaptive capacity to climate-related hazards and natural disasters	**Actively adapt to climate change and strengthen resilience to climate risks in agriculture, forestry, water resources, and other vital fields, as well as cities, coastal regions, and ecologically vulnerable areas.** Gradually establish a forecast, warning, and disaster prevention and reduction system and establish a meteorological early warnings system with full territorial coverage
Goal 15. Life on Land	15.1 By 2020, ensure the conservation, restoration, and sustainable use of terrestrial and inland freshwater ecosystems and their services, in particular forests, wetlands, mountains, and drylands, in line with obligations under international agreements	**Maintain water levels in critical wetlands and estuaries. Protect and restore the ecosystems in wetlands, rivers, and lakes.** Establish a restoration and protection system for wetlands. Establish a legal system for natural land reserves and improve the sustainable access to natural resources, including forests

(continued)

Table 6.2 (continued)

SDGs and targets		Actions taken in China
	15.5 Take urgent and significant action to reduce the degradation of natural habitats, halt the loss of biodiversity, and, by 2020, protect and prevent the extinction of threatened species	By 2020, **establish a substantial network of ecological corridors and biodiversity protection. Launch major biodiversity protection projects. Improve the management of natural reserves, and increase efforts to protect typical ecosystem, species, gene, and landscape diversity**. Increase investment in ecosystem protection and restoration and conduct large-scale surveys on biodiversity. Establish a national monitoring system for biodiversity
	15.9 By 2020, integrate ecosystem and biodiversity values into national and local planning, development processes, poverty reduction strategies and accounts	**Require local governments at all levels to better protect ecological environment and biodiversity according to local conditions** and incorporate the protection measures into local mid- and long-term development plans. Conduct primary survey and **assessment of biodiversity and build a monitoring network and database for biodiversity**. Launch pilot and demonstration projects for biodiversity restoration and protection

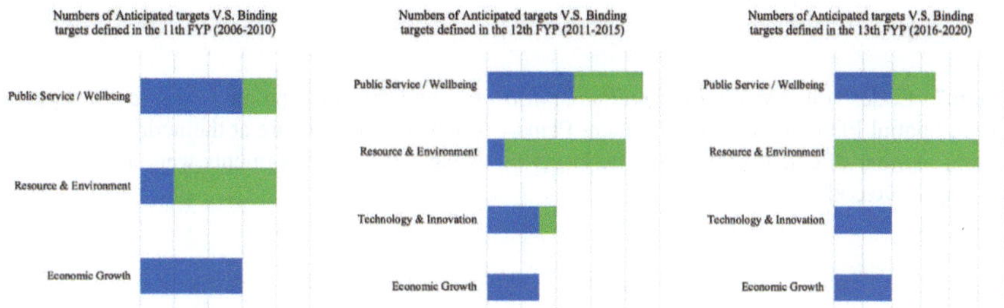

Fig. 6.4 Numbers of anticipated targets *versus* binding targets defined in the FYPs

tendency of shifting from the political philosophy of *Scientific Development* to *Ecological Civilisation*. A slightly different pattern is found in Fig. 6.8, which demonstrates the distribution frequency of contemporary political philosophy **in national institutional reform plans: the term *Ecological Civilisation* gets less mentioned—probably a sign that the institutional reform initiated by the central government around 2003–2005 is moving into a more problem-specific stage requiring greater technical focus**. In comparison, when *Ecological Civilisation* was initially conceptualised in 2003–2005, the term was mentioned much more frequently in the national institutional reform plans.

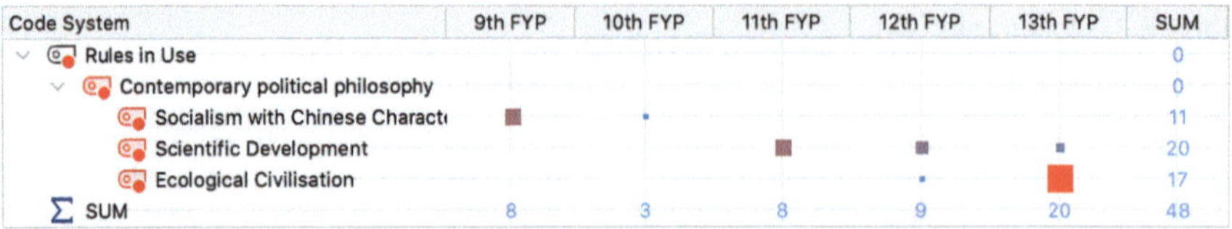

Fig. 6.5 Codes* on *contemporary political philosophy* and their distribution frequency in the FYPs. * Each coloured square at the nodes represents a certain number of coded segments. The larger the square, the more segments were assigned with the respective codes

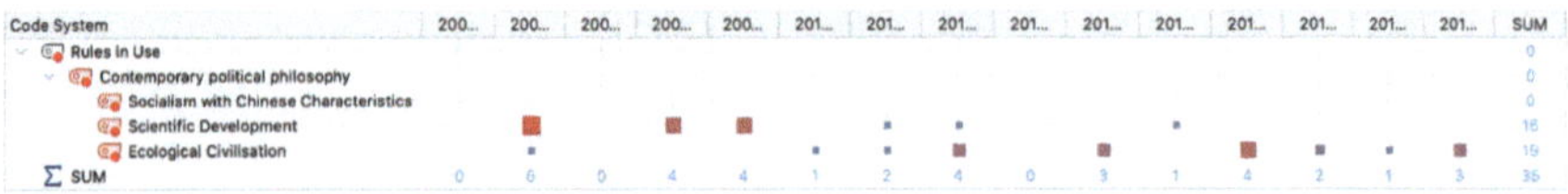

Fig. 6.6 Codes* on *contemporary political philosophy* and their distribution frequency in National Climate Change and Sustainable Development Plans. * Each coloured square at the nodes represents a certain number of coded segments. The larger the square, the more segments were assigned with the respective codes

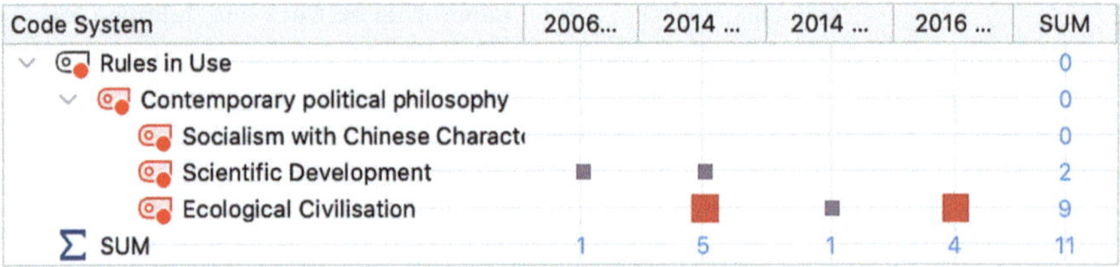

Fig. 6.7 Codes* on *contemporary political philosophy* and their distribution frequency in National Urban Spatial Planning and Development Plans. * Each coloured square at the nodes represents a certain number of coded segments. The larger the square, the more segments were assigned with the respective codes

Fig. 6.8 Codes* on *contemporary political philosophy* and their distribution frequency in National Institutional Reform Plans. * Each coloured square at the nodes represents a certain number of coded segments. The larger the square, the more segments were assigned with the respective codes

6.1.3 Policy Tendencies at the Local Level

This section inspects the policy characteristics and tendencies at the local level through two analyses: (1) changes in the numbers and focus of local development targets in the local FYPs and (2) the distribution frequency of contemporary political philosophy in local plans. These two analyses aim to understand the institutional dynamics further and examine the extent and approaches for urban green space upscaling by the local government, in this case, the Municipal Government of Guangzhou.

Table 6.3 reflects several changes from the 12th FYP of Guangzhou to the 13th FYP, including **an increase in the number of targets on Environmental Liveability, Public Service and Social Well-being, and Infrastructural Upgrade and a decrease in the number of targets on Economic Structure and Growth**. Also, as one of the targets in Environmental Liveability, **"Urban green spaces per capita" changed from an anticipated target to a legally binding one** from the 12th to the most recent 13th local FYPs.

Figure 6.9 shows the distribution frequency of *contemporary political philosophy* in Local Climate Change and Sustainable Development Plans, and Fig. 6.10 shows the distribution frequency of *contemporary political philosophy* in Local Urban Spatial Planning and Development Plans. Contradictory to national-level policy plans, both two figures show that **political philosophy is rarely narrated in local policy plans.** Except in the *2017 Special Plan on Park Construction and Protection*, in which tens statements were made about *Ecological Civilisation*. Whether this number is significant requires further examination.

Table 6.3 Numbers of targets in the 12th and 13th FYPs of Guangzhou

Type of targets	Environmental Liveability*		Public Service and Social Well-being	Economic Structure and Growth	Technology and Innovation	Infrastructural Upgrade
		Urban green spaces per capita				
12th FYP Guangzhou	7	15 $m^2 \rightarrow$ 16.5 m^2 anticipated	11	6	3	5
13th FYP Guangzhou	10	16.5 $m^2 \rightarrow$ 18 m^2 binding	13	4	3	6

Fig. 6.9 Codes* on *contemporary political philosophy* and their distribution frequency in Local Climate Change and Sustainable Development Plans. * Each coloured square at the nodes represents a certain number of coded segments. The larger the square, the more segments were assigned with the respective codes

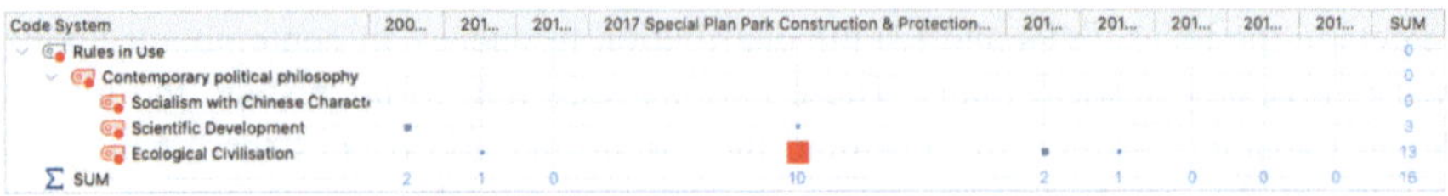

Fig. 6.10 Codes* on *contemporary political philosophy* and their distribution frequency in Local Urban Spatial Planning and Development Plans. * Each coloured square at the nodes represents a certain number of coded segments. The larger the square, the more segments were assigned with the respective codes

6.1.4 Summary

The analyses in Sect. 6.1 shows a steady increase in prioritising sustainable development, low-carbon development, and actions to address climate change impacts at the national level. The increase in the legally binding targets of resource and environment and the decrease in economic growth targets indicate a more robust political will of the Chinese central government to government the environment more effectively and to shift out of the economic growth focus. This shows that the Chinese central government has been aware of the need to address climate change impacts and environmental degradation. There is a tendency to shift contemporary political philosophy in the narratives of national policies and plans from the political philosophy of *Scientific Development* to *Ecological Civilization*, and a recent decline in the frequency of the latter is probably a sign that the institutional reform initiated by the central government around 2003–2005 is moving into a more problem-specific stage requiring greater technical focus.

While national-level policy documents primarily define formal institutional change at the macro and conceptual level, local policy documents mostly describe known costs and benefits in social, environmental, and economic dimensions at the operational level. Besides, the number of technical guidelines has increased from the first to the third case, indicating that the state and local governments are paying greater attention to the technicality and legality of urban green space planning and governance, the increase of which probably has reduced the overall transaction costs to implement these projects. These policy tendencies also indicate that the national and local governments are paying greater attention to the importance of Urban green spaces in urban development. However, contradictory to the national policy plans, political philosophy in local policy plans is rarely narrated. Furthermore, analysis of the local FYPs of Guangzhou shows an increase in the number of targets on Environmental Liveability, Public Service and Social Well-being, and Infrastructural Upgrade and a decrease in the number of targets on Economic Structure and Growth. These findings indicate that local governments focus more on getting the work done and concern less about careful political alignment on paper.

6.2 Relationships and Power Dynamics Between Actors

Following the IAD framework (see 3.5 Analysing Common-Pool Resource Governance through the IAD Framework) but with a small twist, in this research, *communities*[1] in the three Urban green spaces governance cases are treated are equal to the ④ *actors or participants* involved in the three case studies. They include three groups—local governments, communities or local residents (including urban migrants), and landowners, each with its particular attributes. Thus, this section analyses the relationships and nuances of power dynamics between actors in each case study based on the information of revealed in the policy documents and by the interview informants (also revealed in Sect. 5.3 Three Urban green spaces Governance Cases in Guangzhou). The aims are to further understand the institutional dynamics, facilitators, and underlying barriers, to examine the approaches for Urban green spaces upscaling by the local municipal and district government, and to analyse any potential indirect, non-monetary costs and benefits in between actors that influenced the Urban green spaces implementation outcome in the three case studies.

Figure 6.11 shows seven major actors in the three Urban green spaces governance case studies in Guangzhou. The central government is not presented in Fig. 6.11 as it does not participate directly in the three Urban green spaces governance cases. Guangzhou Municipal Government is the highest-level administrative body at the city level, which hosts the three district governments involved in the case studies—Panyu District Government, Haizhu District Government, and Tianhe District Government, respectively. Small businesses include small restaurants, rentals, mechanic services, and small factories. These small businesses in the case studies also contain a degree of land-use informality in the way these businesses were established. Urban migrants are those who came from rural areas outside of Guangzhou and whose household registration, or *Hukou* (in Chinese: 户口; hù kǒu) are rural. Household registrations primarily tied to places of birth, and rural household registration holders can only access the services and welfare of the rural areas. This means that the urban migrants in Guangzhou do not have the same rights as their peers with urban household registration in Guangzhou, in terms of access to urban public services and participation in public affairs. Farmland owners in this research refer to those who collectively own the farmlands and neighbouring village residential land within the urban territory of Guangzhou. Finally, the ordinary residents of Guangzhou possess a general citizen right to monitor and report misdemeanours of government officials, even though they are not directly involved or influenced by the three Urban green spaces planning and implantation cases.

There are four main actors in Case 1—the construction of the Panyu Ecological Corridor network (see Fig. 6.12): Guangzhou Municipal Government, Panyu District Government, small (and informal) businesses, and urban migrants. Guangzhou

[1] The researcher recognises the different connotations that the concept of *community* has in different disciplines. For example, many sociological or anthropological studies emphasise the diversity of individuals within a community and often go into great depth portraying those differences.

Fig. 6.11 Main actors in the three Urban green spaces case study in Guangzhou

Municipal Government and Panyu District Government are **superior and subordinate governments, and they have a relationship of collision, cooperation, and collaboration in social, economic, and environmental protection sectors.** This is also the potential cost between superior and subordinate governments in general. The small (and informal) businesses contribute economic benefits and local GDP targets to Panyu District Government. **Panyu District Government can allow small (and informal) businesses to continue operating or exclude them**. **Migrant workers are "grey" human resources or potential grey economic benefits for Panyu District Government** or any local urban governments, as they do not have to pay for the social welfare of these migrants even though they are the de facto residents. **The relationship between urban migrants and small informal businesses is a mutually beneficial symbiotic relationship between producers and consumers**.

There are six main actors in Case 2—the conversion of the Haizhu Wetland Park from the Wanmu Orchard (see Fig. 6.13): the central government, Guangzhou Municipal Government, Haizhu District Government, small (informal) businesses, urban migrants, and farmland owners who have collective property rights to the farmland. **The central government, Guangzhou Municipal Government, and Panyu District Government are superior and subordinate governments in a collision, cooperation, and collaboration relationship in social, economic, and environmental protection sectors.** This is also the potential transaction costs between superior and subordinate governments. **The small (and informal) businesses contribute economic benefits and local GDP targets to Haizhu District Government, who, on the contrary, can allow the small (and informal) businesses to continue to operate or exclude them**. As the small informal businesses expand, they increasingly erode the farmland in the orchard. Under this circumstance, the transaction costs

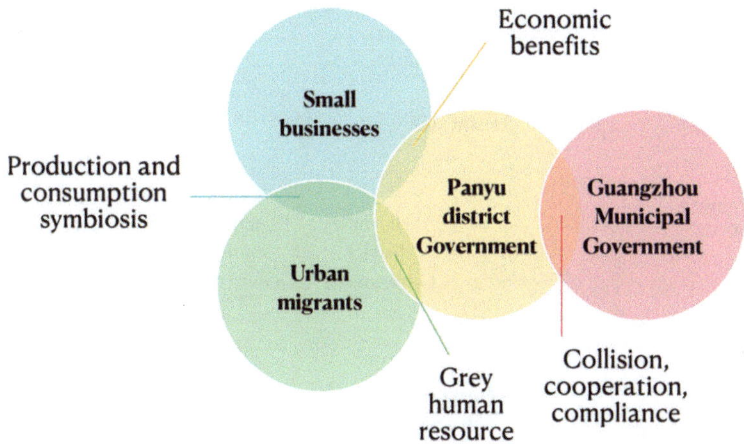

Fig. 6.12 Actor relationships in case 1 Panyu Eco-Corridor network construction

between the Haizhu District Government and the small informal business community gradually increase. **For Haizhu District Government, migrant workers are also "grey" human resources or grey economic benefits**, as the government do not have to pay for the social welfare of these migrants, even though they are also the de facto residents. **Urban migrants and small informal businesses have a mutually beneficial symbiotic relationship as producers and consumers. Urban migrants and farmland owners also have a relationship as producers and consumers, as farmland owners "produce" affordable, semi-formal housing for urban migrants to rent.** But it is not necessarily a mutually beneficial symbiotic relationship. The premise for the symbiosis of the two is that the farmland owners have long-term property rights to the farmland, which became uncertain after the Guangzhou Municipal Government decided to acquire the farmland and transform it into a wetland park.

There are only three main actors in Case 3—the construction of the Tianhe Sponge City demonstration site (see Fig. 6.14): the central government, Guangzhou Municipal Government, and Tianhe District Government. **As superior and subordinate governments, they have a collision, cooperation, and collaboration relationship in social, economic, and environmental protection sectors.** This is also the potential transaction costs between superior and subordinate governments.

6.2.1 Summary

As superior and subordinate governments, the Guangzhou Municipal Government and its cohort district governments have a collision, cooperation, and collaboration relationship in social, economic, and environmental protection sectors. This is also the potential cost between superior and subordinate governments in general.

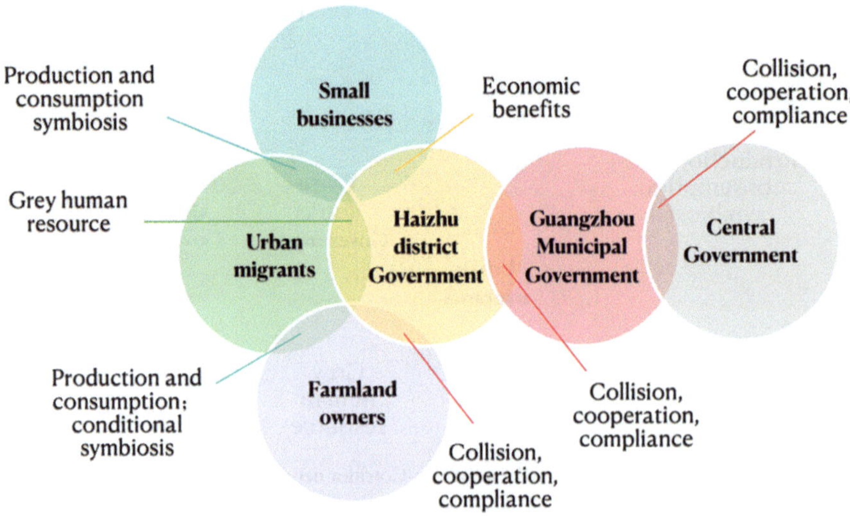

Fig. 6.13 Actor relationships in case 2 Haizhu Wetland Park construction

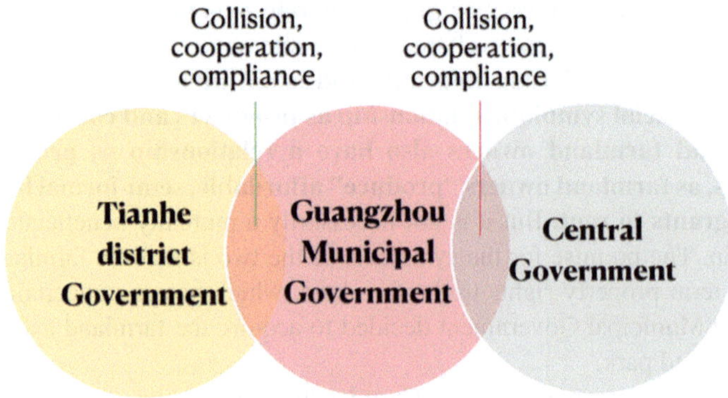

Fig. 6.14 Actor relationships in case 3 Tianhe Sponge City demo-site construction

Small (and informal) businesses benefit the district governments economically and contribute to the local GDP targets. The district governments can allow them to continue operating or exclude them. For local governments, urban migrant workers are "grey" human resources or potential grey economic benefits, as the local government do not have to pay for the public social welfare of these migrants.

Nevertheless, the urban migrants are also de facto residents. The relationship between urban migrants and small informal businesses is a mutually beneficial symbiotic relationship between producers and consumers. Urban green spaces are a common-pool resource (as framed in Sect. 3.4.2) for all these actors—as none of them could be excluded from using it but conserving more urban land for Urban green spaces could hinder the interests, overlapping rights, and responsibilities of all these actors. All these actors are holders of distinctive property rights and the pragmatic "economic man" who seek to maximise their environmental, social, or economic benefits. Urban migrants, as the holder of rural *Hukou* (household registration), do not possess the formal right to participate in urban governance, but ethically speaking, they, as the de facto residents of the city, have an informal right to the same social welfare as others.

6.3 Interactions Between Actors

Based on the relationships and power dynamics between actors analysed in the above section, the actual interactions between actors in each case study are analysed in this section. Figure 6.15 showcases the interactions between all the actors in the three Urban green spaces case studies in Guangzhou, a result of co-occurring the frequencies of the coded segments for actors and interactions in all the in-depth interviews about the three Urban green spaces cases in MaxQDA.

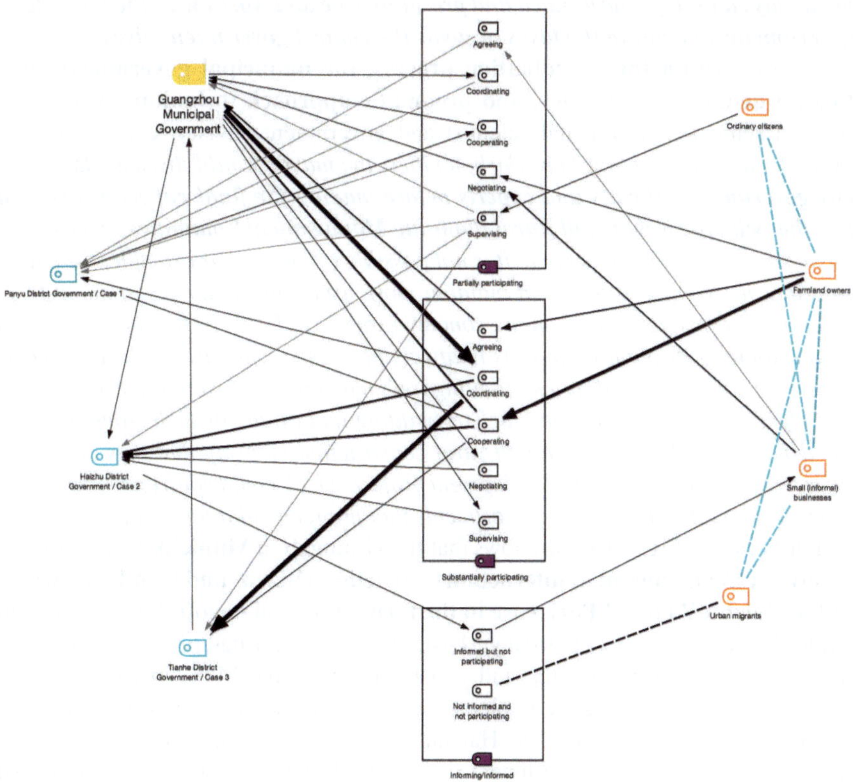

Fig. 6.15 Interactions between actors in the three Urban green spaces case studies

Note: The directions of arrows represent interactions from A to B. For example, ordinary citizens participate partially through supervising the municipal and district governments. The size of the arrows are defined by the co-occurrence frequencies of the coded segments in the chosen in-depth interview transcripts in MaxQDA. The stronger the arrow, the stronger the interaction between two particular codes/actors.

For state actors, the analysis shows that **Guangzhou Municipal Government is the most active actor whose primary role is coordinating with other actors, mainly district governments**. Government advisors also emphasise this primary coordinating function of municipal governments. As revealed in the in-depth interview by Zhao, a senior urban planning researcher and government advisor: "*The common practice of the central government is to give a general task and some general and guiding indicators, for example, building a low-carbon city. How does the local government implement it? Every city needs to plan according to local conditions and development needs. Some cities will focus on the development of green buildings. Some cities may have a better climate, and they would focus on developing more urban green spaces. In contrast, others may focus on optimising the industrial structure, penalising enterprises with high pollution and low efficiency, ordering them to conduct improvements, or ordering them to close. In this process, **the city government plays a highly coordinated and presiding role and works with the lower-level governments to achieve the tasks or goals the central government gives**.*"

Second, **within the coordination process, the municipal government often takes a democratic, creative, and pragmatic approach to building proposals**. Xiao, a senior environmental scientist and government advisor, explained: "*For example, for constructing Low-Carbon cities, the **mayor would discuss plans with city government officials and experts before making the final collective decision together with the Urban and Rural Planning Management Committee**. If the mayor alone says that this plan is good, it is not possible (to pass). For example, the mayor said that he wanted to focus on building more urban green spaces, a department chief on land resources or a land economy expert would likely say that the cost of land resources for planting trees is relatively high and would ask the land economy experts to do some calculations. Planting trees may result in losing some more direct and efficient economic benefits, such as building real estate. In such circumstances, the department chief of urban and rural construction may say that they happen to have a residential real estate development project, which could contribute to building Low-Carbon cities by building apartments that fit green building standards.*"

Furthermore, Fig. 6.15 also shows that **the Guangzhou Municipal Government receives a small amount of interaction from fellow citizens and farmland owners in the Haizhu Wetland Park case in the form of partial negotiation and partial supervision**. For Case 1, the Panyu District Government has primarily interacted passively by partially agreeing and cooperating with the Municipal Government and partially coordinating the implementation of the Panyu Eco-Corridor Network construction plan. For Case 2, the Haizhu District Government has played a similar role as the Panyu District Government, except that the interactions with the Municipal Government have been more intense. And for Case 3, the Tianhe District Government coordinated quite closely with the Guangzhou Municipal Government to implement the Sponge City Initiative.

However, **the conflictual internal coordination dynamics within the municipal governments, particularly in Case 3**, are not evident in Fig. 6.15. He, a senior urban planner who contributed substantially to the *Guangzhou Sponge City Construction Plan*, describes this conflict: "*The Sponge City project has received strong support from the Guangzhou Municipal Government, and now they are establishing a demonstration site in Tianhe District. However, at the same time, the urban planning and construction departments are also experiencing resistance from the Bureau of water resources. As the construction of Sponge City necessitates ecological and natural methods for flood retention and wastewater treatment, there will be fewer budgets in the future for the Bureau of water resources and negatively impact its downstream stakeholders. Therefore,* ***a conflict of interest has arisen between different departments within the municipal government, which can be understood as the coordination or negotiation cost within the municipal government.***"

For non-state actors, first of all, their identities overlap and sometimes have symbiotic relationships (see Figs. 6.12 and 6.13). **Ordinary citizens interact with municipal and district actors through partial supervision**, see an interview excerpt from *Yuan*, a senior urban planner, scholar, and government advisor: "*Ordinary citizens have the legal rights to be informed and make decisions about the environment. Ordinary citizens interact with municipal and district actors through partial supervision. The government generally has a process for publicizing information on construction projects to obtain public feedback on environmental impacts. In addition, the public can participate in improving a particular project through supervision and reporting if they find certain decisions or actions of a government department or an official suspicious or improper.*"

Second, the major actor in Case 2 the transformation of the Wanmu Orchard into the Haizhu Wetland Park—**the farmland owners have had active interactions with the municipal and district governments by partially negotiating**, some of them fully agree with the proposal, and many others cooperated after being informed, without manifesting agreement or disagreement. See the two interview excerpts from Cai, a government official from the Haizhu Wetland Park Management Office: "*In 2011, the Guangzhou Municipal Government submitted to the Ministry of Land Resources the proposal to acquire the Wanmu Orchard farmland without changing the land function. The Haizhu District Government then established a land acquisition work unit with an advocacy team and land acquisition and demolition team.* ***The advocacy team informed the farmland owners with materials, including posters and videos of the farmland acquisition plan, through meetings with question-and-answer sessions with representatives***". "***The representatives of farmland owners, whose role is to express opinions and make decisions for the collective benefits of the farmland owners, have a significant amount of power and long-rooted vested interests in the local social and economic network.*** *The institutional arrangements to trade the farmland ownership for building the Haizhu Wetland Park are against their existing vested interests. Therefore, some farmland owner representatives were* *not enthusiastic about the proposal. They further negotiated with the municipal government on the financial compensation of farmland property right transfer.*"

Third, **urban migrants are the most silent actor of all, with few interactions registered both from them and to them**. See an interview excerpt from *Huang*, a retired government official of the Haizhu District Government: "*The institutional arrangements for the Wanmu Orchard farmland acquisition ignored the impact on some actual residents of urban migrants in surrounding urban villages, intentionally or unintentionally. Transforming the orchard into Haizhu Wetland Park improves the quality of the ecological environment in the area. It is stipulated that 10% of the reserved land would be used for commercial development (hotels, shops,* etc.*) for the farmland owners with Guangzhou urban Hukou. These two interventions jointly increased the semi-informal housing prices in the urban villages. The leasing price also indirectly makes these urban villages more crowded, and the quality of life for those urban migrants who live their further declines.* ***In this process, the migrant workers were indirectly negatively affected by the institutional arrangements for the Haizhu Wetland Park, and they were not officially notified, let alone new institutional arrangements to compensate for the negative impacts.***"

Finally, if the thickest arrows in Fig. 6.15 represent a more ideal level of interaction, then, the thickest arrow can also be interpreted as where two particular actors can interact with least cost. Then, some further assumptions may be made about the thinner arrows and the greater costs between those actors. This will be discussed in greater detail in Chap. VII.

6.3.1 Summary

In the three urban green space planning and governance case studies, the Guangzhou Municipal Government is the most active actor whose primary role is coordinating with other actors, mainly district governments. Within the coordination process, the municipal government often takes a democratic, creative, and pragmatic approach to build proposals with each department and urban planning experts. In this process, the primary costs taken into consideration by the municipal government are budget, technical expertise, time to reach a political consensus between departments, and time to implement these Urban green spaces plans. The technical expertise includes expertise in ecological planning and the arrangements of rights among actors to access, manage, withdraw, and exclude. Citizens often interact with the Guangzhou Municipal Government through information solicitation and supervision. Citizens often interact with the Guangzhou Municipal Government through information solicitation and supervision. In this process, the primary costs bared by citizens include the time of involvement and difficulty accessing information about these projects. Farmland owners can, and some have interacted with, the municipal and district governments through negotiation and supervision. In this process, the primary costs bared by the farmland owners are:

- risks of losing the rights to withdraw and manage the orchards,
- losing local agricultural traditions for the next generations,
- not getting what they considered decent compensation from the local governments.

In general, the non-state actors' identities overlap and, in some cases, have symbiotic relationships. There are also costs involved in maintaining these symbiotic relationships—as a group of non-state actors, their transaction costs facing the governmental actors are based on their weigh-in on individual interests *vs.* collective interests in these Urban green spaces planning and governance projects.

6.4 Outcome in Urban Green Space Implementation and Institutional Change

This section analyses ⑦ *outcomes* in the three Urban green spaces governance case studies, including the goals and deliverables of Urban green spaces planning/implementation, and resultant formal and informal institutional change.

6.4.1 Urban Green Space Goals and Deliverables from National and Local Plans and Policies

This section presents five principal analyses related to the outcome of urban green space planning and governance in Chinese cities, in general, and in the three case studies, during the study period. These analyses were made around (1) the Urban green spaces actions or deliverables from the national- and local-level Urban green spaces-related climate plans or spatial planning policies, (2) the reflections on Urban green spaces implementation with evidence from the in-depth interviews, (3) the formal institutional changes which reveal evolving government priorities and local-level institutional changes as a response to central-level institutional changes, (4) the reflections on institutional change with evidence from the in-depth interviews, and lastly, (5) the types of informal institutional changes incurred. The analysis aims to examine to which extent local governments have been able to advance and upscale Urban green spaces and with what approaches.

Figure 6.16 shows a series of Urban green spaces interventions considered in official plans for urban climate change adaptation and biodiversity conservation. These interventions include urban surface coverage, urban green vegetation as a carbon sink, wild plants and species genetic pool protection, roof gardening, rain gardens, biological water retentions, and more. It also shows that **the linkage between Urban green spaces and urban climate change mitigation and adaptation in China has constantly been present in the national policy documents from 2007 to 2019**.

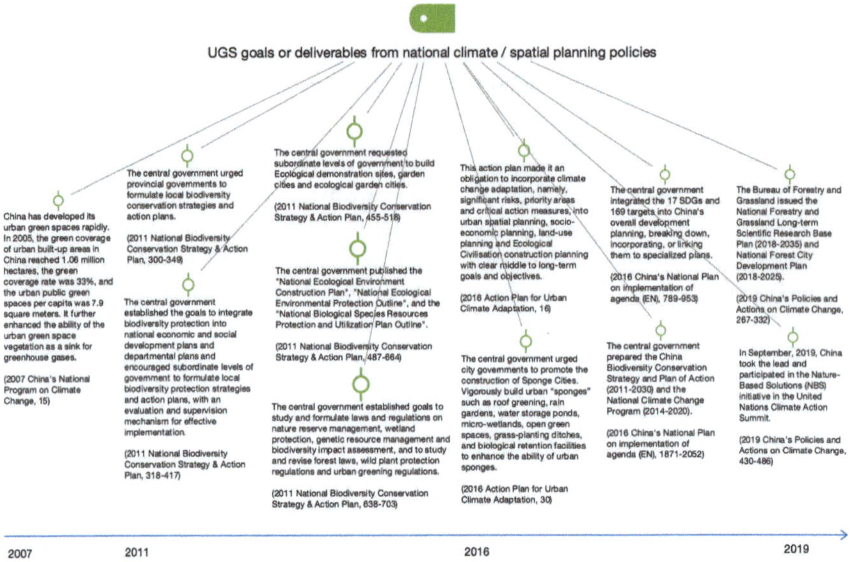

Fig. 6.16 Urban green spaces goals and deliverables from the national Urban green spaces-related climate plans or spatial planning policies during the case study period

Then, information on Urban green spaces intervention in the local Urban green spaces-related climate change and urban planning plans and policies during the case study period is extracted and presented in Fig. 6.17. It shows **the ambitious goals of the Municipal Government of Guangzhou in increasing the size of Urban green spaces per capita and the rate of total urban surface coverage, as well as strict land-use control and ecological protection**. Furthermore, it shows that **the municipal government laid out concrete measures to build Sponge City to alleviate urban pollution and inundation, including adding vertical greens, recessive green spaces, wetlands, and rainwater gardens, and further emphasised institutional support, including policies, regulations, specifications, and evaluation standards to help it succeed**. Notably, even when these urban green space action items or goals address climate change's impacts, such as urban inundation, these local-level plans did not link them directly to the word "climate change." Fig. 6.17 also showcases a linkage between Urban green spaces interventions and their role as a public service. The policy document analysis above shows that **the Municipal Government of Guangzhou responded positively to the Urban green spaces goals and requirements put forward by the central government**. Historical data on urban green space implementation at the national and local levels also suggested a positive correlation.

Figure 6.18 shows a similar increasing trend in both the percentage of Urban green spaces size *vs.* all urban lands and the sizes of Urban green spaces per capita, for Guangzhou and national average, respectively. **Particularly between 2005 and 2018, Guangzhou had an above-national-average increase in both Urban green spaces size per all urban lands and per capita**.

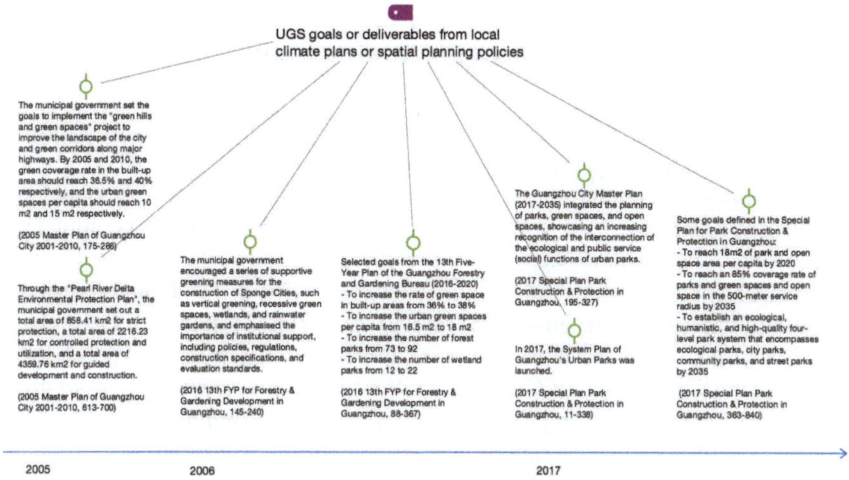

Fig. 6.17 Urban green spaces actions or deliverables from the local Urban green spaces-related climate plans or spatial planning policies during the case study period

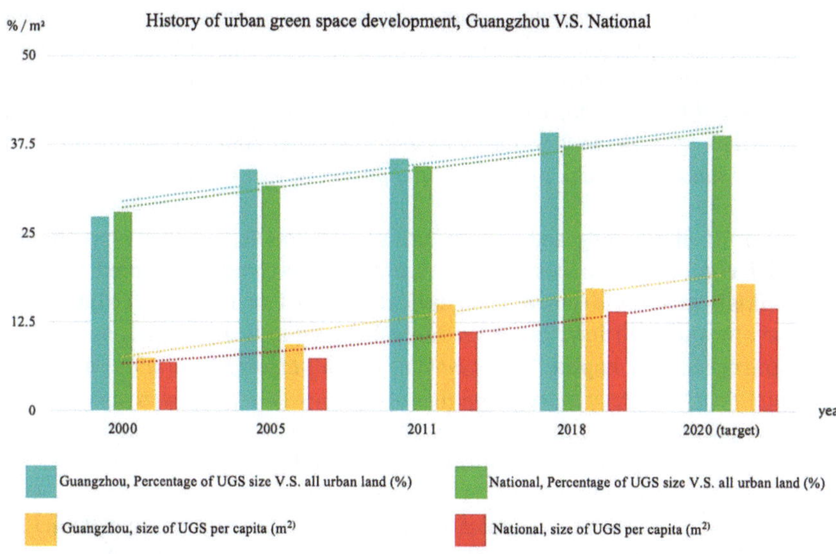

Fig. 6.18 Linear increase trend in urban green space development, Guangzhou versus National in selected years. *Data source* national and local government reports

6.4.2 *Reflections on Urban Green Space Implementation—Insights from In-depth Interviews*

This section presents selected excerpts from the in-depth interviews with 50 professionals of urban planning, urban governance, environmental science, land and political economics, and public administration. These excerpts cover general knowledge as well as specific experiences on Urban green spaces policies and intervention from these professionals.

Xiao, a senior environmental scientist, argues that **China's Urban green spaces planning practice has become more ecology-oriented compared to the conventional approach and describes the stark land-use tension in the recent rapid urban expansion history in Guangzhou**: "*In China, the traditional urban green space planning practice is to plan urban green spaces as a supporting facility, that is, to provide public services. Similar to schools and medical institutions, the amount and distribution of Urban green spaces planned are based on the calculation of the scale and density of the residential population. Due to the technical difficulty of quantifying its direct economic benefits, urban green space planning in China has not been deliberately oriented toward economic services. But urban green spaces may improve the quality of a residential community, thereby increasing property value. Therefore, most urban green spaces are planned as public services and supporting facilities. But now, **with the popularisation of concepts such as environmental protection and Ecological Civilisation, Urban green spaces planning is slowly considering the ecosystem services**, that is, adopting the ideas and concepts similar to those proposed by Ian McHarg and protecting land properties with exceptional ecological functions.*"

Rao, an urban planning and architecture scholar, points out that **even though national Urban green spaces policy orientation has gradually opted towards more sophisticated requirements and indicators, most current local Urban green spaces evaluation still focuses on incomprehensive indicators such as "Urban green spaces size per capita:"** "*The current urban planning system has increased the requirements for green space planning greatly. However, most current evaluation of green space quality only focuses on the coverage area or green space per capita. The ecological concepts and indicators of species and biodiversity are still relatively rare.*"

Yue, a researcher in land economics and urbanisation, reveals that **private actors also participate in creating more Urban green spaces but often through commercial means, which often create gentrification**: "*Since the direct economic benefits of Urban green spaces are difficult to estimate, for economic efficiency considerations, the government usually packs/bundles Urban green spaces into a commercial complex, such as a commercial housing complex, to increase the amount of Urban green spaces in the city.*"

Yu, a senior urban landscape architect, scholar, and practitioner, criticises the dubious city beautification movement and the negative impact of the hardening of natural rivers and **considers the Sponge City demonstration site in Guangzhou as**

a novel and nature-based design more effective in treating urban pollution and climate change impacts: "*It is sad to see that, in the process of rapid urban expansion, the canalization and hardening of natural rivers and the "beautification" movement are continuing to increase in large and small cities in China. We can be wiser in how we utilize our rivers and natural landscape. The Tianhe Daguan Wetland Park, as a pilot project of Guangzhou's Sponge City construction, for example, demonstrates the possibility of a novel design for urban greening and landscaping. In this case, we used the natural resources at the original site and the least design to create a genuinely economical urban green space, provide contemporary urban residents with the most and best ecological services, slow down urban waterlogging, and treat water pollution ecologically. It integrates the needs of residents, management of environmental pollution and climate change impacts, and preserving native plant species.*"

Lv, a senior government official of the urban planning system of Guangzhou, **reaffirms the overall improvement in the city's urban environmental quality and the potential for improvement in Urban green spaces governance**: "*The environmental quality of Guangzhou needs to be improved. The total amount of urban green spaces has increased, but so is the trend of ecological landscape fragmentation, While the proportion of urban construction land has decreased. Although Guangzhou is in the subtropical zone with abundant rainfall, the city has faced a lack of usable, quality water sources due to the severe pollution of rivers.*"

Huang, a retired senior government official in the Haizhu District Government, approves the **positive change in the ecological condition of the Haizhu Wetland Park and the overall environmental quality, admitting a few unintended externalities, including the circumstance of the urban migrants**: "*Farmland owners have lost the right to manage the orchard, running the possibility of losing their agricultural traditions and knowledge not being able to be passed on to the next generation. And because farmland owners could no longer develop their farmland into informal housing for urban migrants, many have made informal additions to their existing homesteads, which surpasses the legal height limit for these homesteads. As a result, the urban villages surrounding the Haizhu Wetland Park have become more crowded, and the living environment of urban migrants has further deteriorated.*"

6.4.3 Formal Institutional Changes, Evolving Priorities, and Central-Local Correlation

This section analyses the formal and informal institutional changes accompanying the Urban green spaces interventions revealed in Sects. 6.4.1 and 6.4.2. The aim is to understand further the changes in the institutional dynamics between actors and between the central and local levels and to examine to what extent local-level governments responded to central-level government objectives.

Figure 6.19 shows the formal institutional changes outlined in the 11th, 12th, and 13th national social and economic development Five-Year Plans (FYPs), respectively, and the occurrence of the three Urban green spaces planning cases along the same timeline. Formal institutional change refers to the written modifications in formal policies, plans, laws, and regulations by government authorities. **The most notable trend is the increase in the requirements for formal institutional changes in environmental protection and climate change mitigation and adaptation** (text in green). Besides, there was only one formal institutional change requirement related to the economic system reform (text in red), and it was in the 11th FYP (2006–2010). Furthermore, **an equal amount of formal institutional change requirements is observed for structural or functional government internal reform** (text in blue) **and environmental protection and climate change mitigation and adaptation** (text in green). A slight increase in the requirements of institutional reform for urban development and urban–rural integration (text in purple) is perceived. During the 11th and the 12th FYP (2006–2015), formal reforms to increase the roles of multiple actors in governance (text in brown) were also outlined. To sum up, Fig. 6.19 shows evolving priorities for formal institutional change at the national level, including decreasing focus on economic system reform, increasing focus on structural or functional government internal reform, and particularly evident increasing focus on environmental protection and climate change mitigation and adaptation.

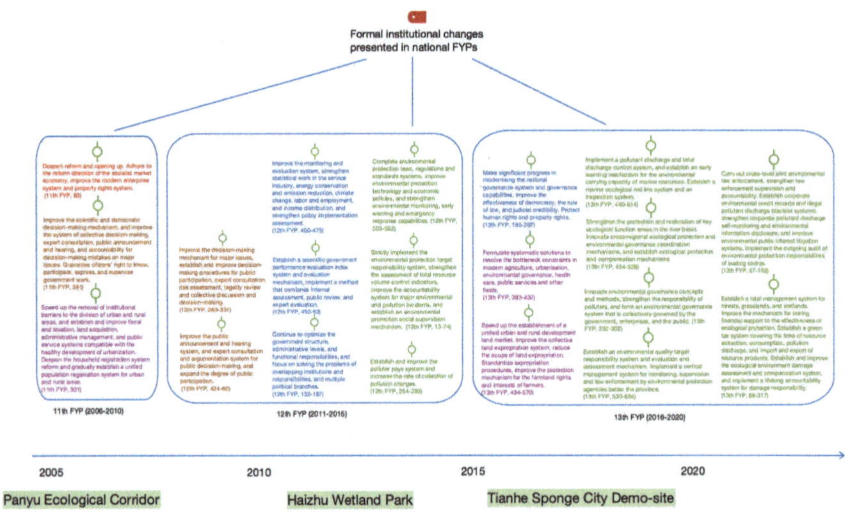

Fig. 6.19 Formal institutional changes introduced in the national FYPs. *Text in red:* formal institutional changes related to the economic system. *Text in brown:* formal institutional changes related to governance and roles of multiple actors. *Text in blue:* formal institutional changes related to government internal reform, structural or functional. *Text in green:* formal institutional changes related to environmental protection and climate change mitigation and adaptation. *Text in purple*: reform for urban development and urban–rural integration

An examination through the national-level climate plans and spatial planning policies finds **four major areas of formal institutional change: (1) environmental regulations, (2) government reform, (3) economic system reform, and (4) the role of societal actors and governance** (see Fig. 6.20). Half of the formal institutional changes outlined in these policy documents focus on government reform, ¼ on environmental regulations, and less than 10% on economic system reform or the role of societal actors and governance. It is also notable that more institutional changes have been planned during the transition years between the national FYPs or the macro socio-economic planning cycles, for example, in 2005 (between the 10th and 11th FYPs) and 2014–2016 (between the 12th and 13th FYPs).

Having examined how formal institutional change has been outlined in official policy documents at the national and local levels, respectively, it is necessary to assess how local-level governments, in this case, the Municipal Government of Guangzhou, has reacted to the national-level formal institutional changes led to the correlations between central and local levels. As shown in Fig. 6.21, **local-level institutional changes primarily incurred/responded to national-level institutional change in environmental protection and climate change mitigation and adaptation** (text in green). Less than one-third focus on internal government reform. Only a tiny amount focuses on urban development, urban–rural integration or governance, and the roles of multiple societal actors. The call to establish a government performance index and to implement a pollutant discharge system is the most responded institutional change item from the local level. **In general, institutional changes at the local level have**

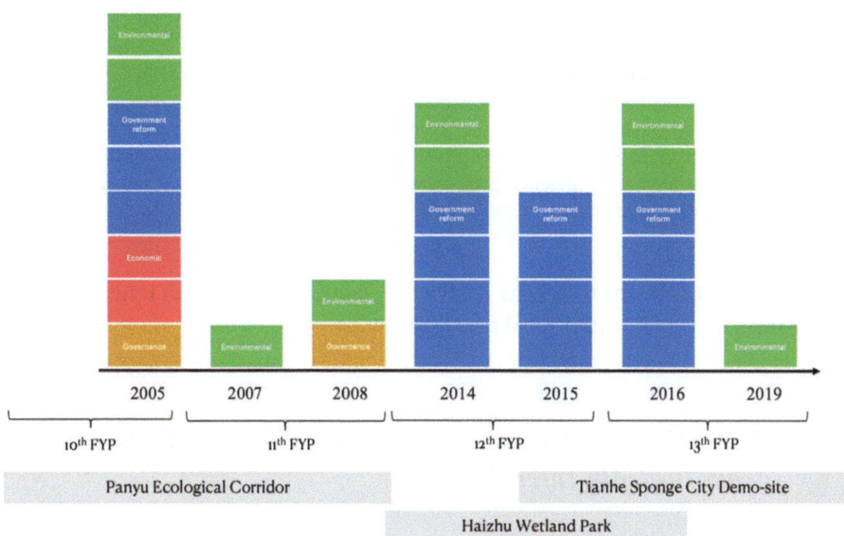

Fig. 6.20 Areas and numbers of formal institutional changes presented in national climate plans and spatial planning policies. *Green*: environmental regulations. *Blue*: government reform. *Red*: economic system reform. *Brown*: the role of societal actors and governance

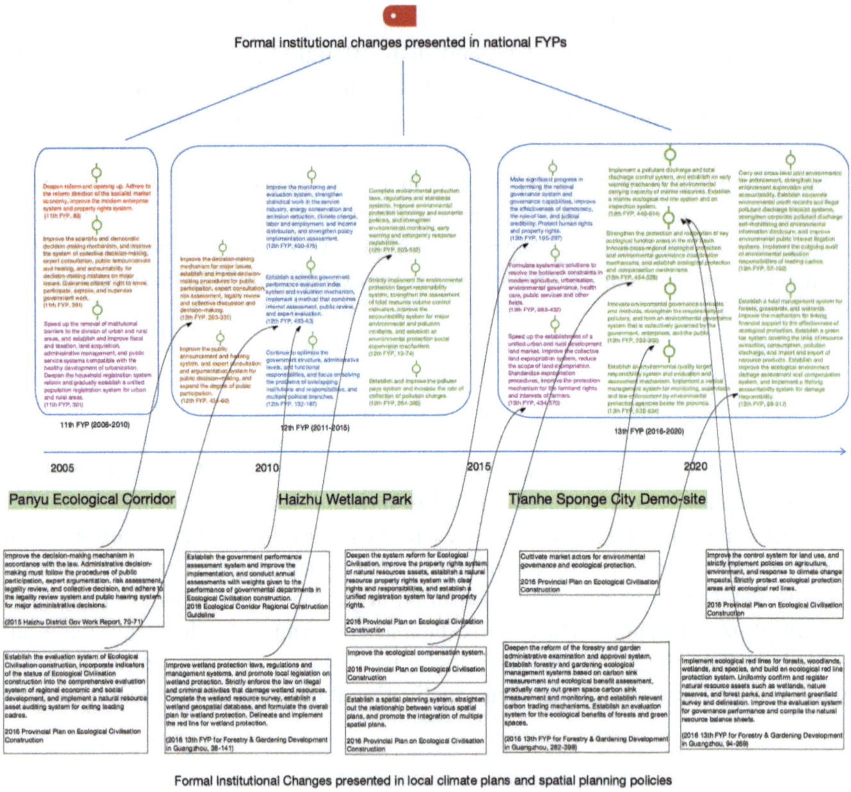

Fig. 6.21 Correlation between local and national levels in formal institutional change. *Text in red:* formal institutional changes related to the economic system. *Text in brown:* formal institutional changes related to governance and roles of multiple societal actors. *Text in blue:* formal institutional changes related government internal reform, structural or functional. *Test in green:* formal institutional changes related to environmental protection and climate change mitigation and adaptation. *Text in purple:* reform for urban development and urban–rural integration

reacted more to the more recent FYPs, for example, the 13th FYP (2016–2020), indicating more vital central-local coordination.

6.4.4 Reflections on Formal Institutional Change—Insights from In-depth Interviews

Reflections from the interviewees can support one of the aims of this research to identify and understand intrinsic value-related factors of potential institutional barriers. The reflections from the interviewees on formal institutional change mainly focus on four aspects: (1) the introduction of ecological performance measures, (2) the unification and systematisation of urban planning, (3) property rights reforms, and (4) government reform with regards to rule of law, accountability, and participation.

One aspect of formal institutional change is the ecologicalisation of evaluating and promoting government officials' performance through environmental protection indicators and natural resource asset audit sheets. Zhao, a senior urban planning researcher and government advisor, argues that this aspect can strongly motivate the government to deliver environmental goals: "*Environmental protection is now the most important indicator to evaluate the performance of local governments: once any major environmental, ecological, and environmental pollution problems have occurred, they will not be promoted to higher positions. This provides local governments with a stronger motivation and passion for environmental protection and management. They are now making corresponding adjustments to their local development policies more actively, focusing on environmental protection. Previously, the environmental protection work was only driven by political pressure, but today it is driven by a more formal governance system with accountability mechanisms.*"

Another aspect is **the unification and systematisation of urban planning with a reduction in the types of plans and departments in charge but with increased land-use restrictions and regulations to protect the environment**. Zhao, senior urban planning researcher and government advisor, calls this change the "modernisation of governance capabilities:" "*One of the actions was establishing the Ministry of Natural Resources to manage various natural resources in a more unified and systematic manner. The biggest benefit of this reform action was to enable more efficient and economical use of natural resources, which is a major manifestation of Ecological Civilisation... In the past, the ecological red line was rarely considered when making an urban plan. Now, full attention to ecological and environmental protection is required. As urban planners, we need to follow the "rules of the three lines" strictly. These three strictly controlled lines are ecological space protection, ecological red, and basic farmland protection lines. Secondly, the three lines in the previous plan belonged to three different departments: The Ministry of Housing and Urban–Rural Development, the Ministry of Land Resources, and the Ministry of Agriculture, respectively. We need to integrate all these lines in any determined urban plan.*"

Additionally, **institutions around property rights arrangements have become increasingly complex**. *Zhao*, a senior urban planning researcher and government advisor, gave an example of collectively owned farmland in rural areas: "*For example, collective land in rural areas could have the same functional attributes as before (the reform of land property rights), but now it's managed uniformly throughout the country. E.g., a plot of collectively owned farmland is now defined as a key ecological protection area; for this reason, although farmers continue to own the farmland collectively, they can no longer develop it for farming or other purposes. Their collective ownership remains for farmers, but their income has changed from the original development and use model to the current ecological conservation model.*" At this time, the local government should compensate the farmers financially for replacing their source of livelihood with ecological conservation.

Furthermore, **a series of reforms in the government system tends to create more political space for societal actors and strengthen the rules of the law**. Li, a senior urban development expert and government advisor, revealed how to go about modernising the government and urban governance: "*(I think that we would focus on) the reform of the administrative management system: transform government functions, reduce administrative levels, improve administrative efficiency, form good government-enterprise and government-society relations, thereby modernizing urban governance. The result of the reform should reflect the requirements for the integrated development of urban and rural areas, greater autonomy of the private sector and society, and the spirit of the rule of law. At the same time, reform the cadre appointment management system: change the traditional economic development-based performance model, instead, incorporate livelihood improvement Ecological Civilisation construction into the cadre evaluation system.*"

All these four formal institutional change trends have been advanced through small-scale trial-and-error experiments and from the central to the local level. Yuan, a senior urban planning researcher, practitioner, and government advisor, shares his insights: "*Many existing types of urban plans in China are a result of a series of trials and errors by selected local governments. The experiments and experiences are gradually summarized, assessed, and approved by the central government before they were institutionalized and promoted nationwide.*"

6.4.5 Informal Institutional Changes

This section presents five major types of informal institutional changes identified in all the national- and local-level policy documents as well as in-depth interviews. The aim of this analysis is to further understand the institutional dynamics, approaches, and emerging norms or intrinsic value-related factors which influenced Urban green spaces governance in China. **These five major types of informal institutional changes are: (1) enhanced urban–rural coordination in urban planning, (2) enhanced notion of public participation, (3) enhanced ecological notion in urban planning, (4) increase in government officials' environmental awareness, and (5) enhanced citizens role for government supervision** (see Fig. 6.22).

6.4.6 Summary

Analysis of urban green space goals in national policy documents finds a series of Urban green spaces interventions outlined for urban climate change adaptation and biodiversity conservation, including urban surface coverage, urban green vegetation as a carbon sink, protection of wild plants and species genetic pool, and more. It also shows that the potential of Urban green spaces for urban climate change mitigation and adaptation is believed and has constantly been considered by the Chinese national

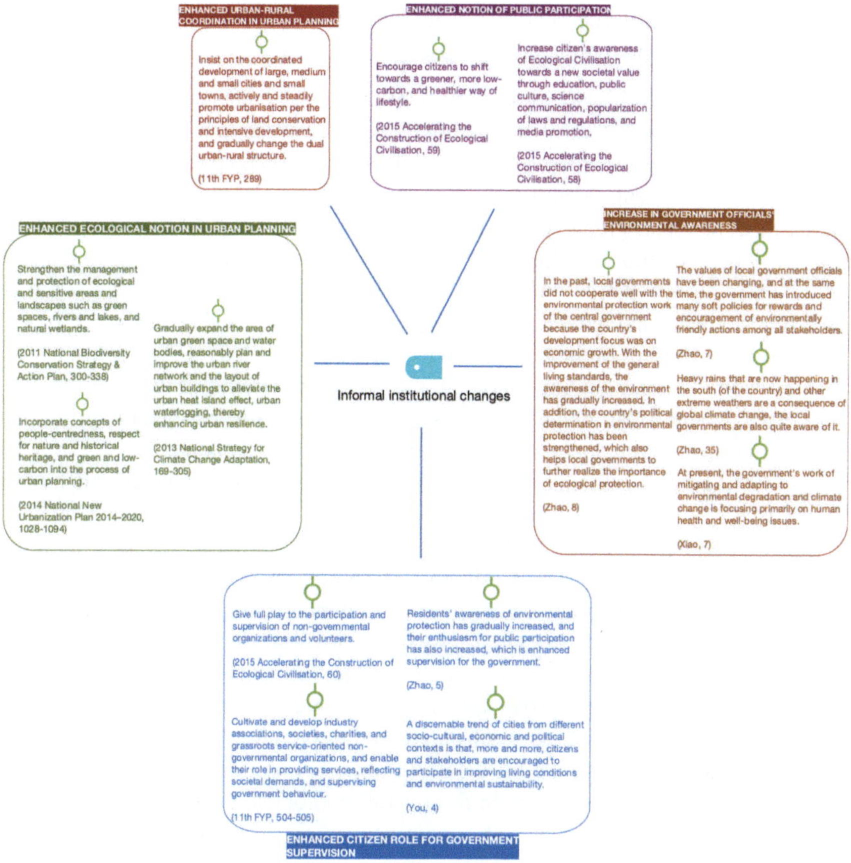

Fig. 6.22 Five types of informal institutional changes identified in policy documents and in-depth interviews

government. Analysis of local policy documents shows the Guangzhou Municipal Government's ambitious goals and concrete measures to increase the Urban green spaces size per capita and per total urban surface coverage and enhance land-use control and ecological protection. The analysis also shows the local government's intention to strengthen institutional support, including policies, regulations, specifications, and evaluation standards, to help it succeed, indicating a positive sign of systematic thinking. Notably, even when these Urban green spaces action items address climate change's impacts, these local-level plans did not link them directly to "climate change." There are three possible reasons:

1. a perception gap of climate change between the central and local governments,
2. the technical-scientific challenge is to confirm the direct causal relations between local economic activities and regional climate change impacts or to quantify the benefits of Urban green spaces ecosystem services at local levels appropriately,

3. a strategy with differentiated focus distributed across levels of government and scales of time, population, and resources.

Comparing the actual historical development of urban green space implementation at the local levels with that of the national average reveals a positive response by the Guangzhou Municipal Government to the urban green space goals defined by the central government. Traditional urban green space planning practice in China tended to treat urban green spaces merely as a product for public service but has evolved to become more ecology-oriented in recent years. The Urban green spaces policy orientation at the national level has gradually opted towards more sophisticated requirements and indicators. However, most current local Urban green spaces evaluation still focuses on incomprehensive indicators, such as "urban green spaces size per capita." These findings indicate the ongoing learning progress of government officials and urban planning professionals, with obstacles remaining in operationalising those more sophisticated requirements and indicators. These obstacles require substantial investment in technical expertise, institutional change, and human resources.

Due to high-speed urban expansion, Guangzhou has suffered stark land-use tension in recent decades. The hardening of natural rivers and land surfaces has caused many negative impacts on urban environmental health. Besides, many urban greening attempts along the so-called "city beautification" movement did not resolve the problems. Furthermore, even though private actors have been allowed and encouraged to create more Urban green spaces via commercials, they have often produced externalities such as gentrification. Interventions like the second case (Haizhu Wetland Park) and the third case (Tianhe Sponge City demonstration site) have resulted in positive changes in the overall ecological condition and strengthened climate resilience. However, shortcomings remain. These reflections show that a more systemic approach is needed to improve the technical expertise and institutional support and include more non-state actors, specifically those whose property rights are informal or poorly established.

Structural or functional government internal reform, environmental protection, and climate change actions are equally essential priorities for the central government. Analysis of the formal institutional change outlined in national policy documents finds a notable increase in the requirements of institutional change environmental protection and climate change mitigation and adaptation and a drastic decrease in the requirements of economic system reform. These findings indicate a sharp rise in the central government´s political determination to tackle environmental and climate challenges. Analysis of the formal institutional change in local policy documents finds more institutional changes being planned during the transition years between the national FYP planning cycles. Most of those focus on government reform and environmental protection and climate change actions, showcasing positive responses to the institutional changes outlined at the national level and mainly responding to the more recent FYPs, indicating stronger multi-level coordination.

Reflections from the interviewees on formal institutional change mainly focus on four trends: (1) the ecologicalisation in the evaluation and promotion of government officials' performance through environmental protection indicators and natural resource asset audit sheets, (2) the unification and systematisation of urban planning with a reduction in the types of plans and the numbers of departments in charge, but with an increase in land-use restrictions and environmental protection regulations, (3) increasing complexity in property rights arrangements, and (4) government reform in critical aspects, including societal actors and the rules of the law. These four formal institutional change trends have been advanced through small-scale trial-and-error experiments, first at the local levels and, once proven effective, promoted by the central government and scaled up nationwide. These findings show great efforts of the central and local governments in institutionalising mechanisms of self-check regarding environmental deeds; in the meantime, the rights or the legal power to govern the environment and climate change are more concentrated in fewer departments.

Furthermore, five major types of informal institutional changes are identified:

1. enhanced urban–rural coordination in urban planning,
2. the enhanced notion of public participation,
3. the enhanced ecological notion in urban planning,
4. increase in government officials' environmental awareness,
5. enhanced citizens' role for government supervision.

These findings show changing values of governments in urban environmental governance, with (1) greater attention to improving the government's environmental knowledge and awareness, which is also helpful in reducing the costs of environmental damage or mistakes made by improper response measures and (2) mobilisation of non-state actors to supervise the government, which could help reduce the costs of law enforcement and internal government supervision.

6.5 Final Evaluation Through Multi-criteria Assessment (MCA)

The above analyses have teased out many subtleties about the ⑥ *interactions* and conflicts between the ④ *actors* in the three case studies, the ⑦ *outcome* in Urban green spaces implementation as well as formal and informal institutional changes. This section evaluates the ⑥ *interactions* and ⑦ *outcomes* through a comprehensive multi-criteria assessment (MCA), as guided by the IAD framework (see Sect. 3.5 Analysing Common-Pool Resource Governance through the IAD Framework, and Fig. 4.9: The IAD framework applied in this research). The MCA is conducted/scored by the researcher according to the information provided by interviewees/informants as well as the analysis done in the above sections. As introduced in Sect. 3.5, considering the flexible and context-specific nature of the IAD framework, this research extends the five basic evaluative criteria that Ostrom (2005) defined into a more comprehensive set of criteria (see Table 6.4), based on these three additional references: (1) the universal principles for effective governance in the *Millennium Ecosystem Assessment (MEA)*, (2) the *Technical Assessment Synopsis Reports* of the European Green Capital Award, and (3) the multi-criteria evaluation (MCE) guideline for ENavi—the Energy Transition Navigation System developed to evaluate the German national energy transition scheme "Energiewende." For reasons of selection of these references, see Sect. 4.3.2.2. The new set of evaluation criteria enables a more detailed evaluation of the variables, thereby increasing the credibility of the results of this MCA.

As showcased in Table 6.4, the first case—the Panyu Ecological Corridor construction project is quite successful in terms of temporal efficiency of planning and inclusion of economic benefits. It also scores high in economic predictability, innovativeness in governance, and transferability of practices. The Panyu Ecological Corridor construction project scores low in social equity, technical/ecological adaptability, resilience, and robustness, accountability, and conformance to general morality, particularly in information transparency and legitimacy (conformity to law).

The second case—the Haizhu Wetland Park transformation project is highly successful in general, particularly in the criteria of economic efficiency in economic, technical, and institutional dimensions, as well as in the criteria of adaptability, resilience, and robustness in technical/ecological, economic, and institutional dimensions, and in institutional accountability. The Haizhu Wetland Park transformation project scores slightly lower in social equity and conformance to general morality.

The third case—the Tianhe Sponge City demonstration site construction is deemed successful in general, particularly in the criteria of adaptability, resilience, and robustness, accountability in technical/ecological and economic dimensions. This case has also demonstrated high efficiency in planning and high compliance with upper-level governments and agendas. Overall, the local government, in this case, the Guangzhou Municipal Government and the Tianhe District Government, did well in fulfilling administrative accountability within own levels and have been quite transparent in terms of public information. The project scores high in legitimacy (conformity to the law) and ethical acceptability. It is comparatively less successful in the inclusion of public participation and there is no sufficient information yet to determine the institutional adaptability over time.

The commonalities shared by all these three cases are the evident temporal efficiency in planning and innovativeness in governance in each own terms, despite not all leading to successful Urban green spaces implementation outcomes. The second and third cases—the Haizhu Wetland Park and the Tianhe Sponge City demonstration site projects, respectively—share high technical/ ecological adaptability, resilience, and robustness.

Table 6.4 Final evaluation of Urban green spaces planning and governance in Guangzhou through a multi-criteria assessment (MCA)

First tier and structural criteria	Dimensions	Second tier criteria	First case	Second case	Third case	Corresponding subcodes (subthemes)	Corresponding categories
1. Economic efficiency	Economic	1. Financing adequacy	N/A	7	N/A	Urban green spaces goals from plans/policies Urban green spaces implementation	⑦ **Outcome**
		2. Direct cost–benefit efficiency	N/A	7	N/A		
	Technical	3. Temporal efficiency of planning	7	7	7	Urban green spaces goals from plans/policies Urban green spaces implementation	⑦ **Outcome**
		4. Temporal efficiency of implementation	1	7	7		
	Institutional	5. Interlinkages between institutions	2	6	5	Urban green spaces goals from plans/policies Urban green spaces implementation Urban climate governance	⑦ **Outcome**
2. Equity	Social	6. Cultivation of social cohesion	2	5	4	Urban green spaces implementation Informing/informed Partially participating Substantially participating	⑦ **Outcome** ⑥ **Interactions**
		7. Inclusion of health and well-being benefits	N/A	5	6	Human health and well-being	⑦ **Outcome**
		8. Inclusion of rights to access information	2	6	7	Urban green spaces implementation Informing/informed Partially participating Substantially participating	⑦ **Outcome** ⑥ **Interactions**
		9. Inclusion of rights to monitor	1	6	6		
		10. Inclusion of rights to participate	2	5	4		

(continued)

Table 6.4 (continued)

First tier and structural criteria	Dimensions	Second tier criteria	First case	Second case	Third case	Corresponding subcodes (subthemes)	Corresponding categories
	Ecological	11. Environmental protection and resource conservation	4	7	7	Urban green spaces implementation Climate resilience	⑦ **Outcome**
	Economic	12. Inclusion of economic benefits	6	5	5	Urban green spaces goals from plans/policies Urban green spaces implementation	⑦ **Outcome**
3. Adaptability, resilience, and robustness	Technical/ ecological	13. Sustainability in land-use	3	6	6	Urban green spaces implementation Climate resilience Human health and well-being	⑦ **Outcome**
		14. Environmental quality (air, noise, waste, water)	2	7	7		
		15. Ecological resilience (ability to stay resilient)	N/A	7	7		
		16. Effectiveness in climate change mitigation and adaptation	4	7	7		
		17. Significance and relevance to the SDGs	4	7	7		
		18. Technical replicability	2	7	7		
	Economic	19. Economic predictability and contribution to the common good	6	6	7	Urban green spaces goals from plans/policies	⑦ **Outcome**
		20. Green growth and eco-innovation	5	6	5		

(continued)

Table 6.4 (continued)

First tier and structural criteria	Dimensions	Second tier criteria	First case	Second case	Third case	Corresponding subcodes (subthemes)	Corresponding categories
	Institutional	21. Innovativeness in governance	7	7	6	Urban green spaces goals from plans/policies Urban green spaces implementation Formal changes Informal changes Urban climate governance	⑦ **Outcome**
		22. Institutional adaptability (changes over time)	5	6	N/A		
		23. Transferability of practices	6	6	6		
4. Accountability	Institutional	24. Compliance with upper-level governments and agendas	1	7	7	Urban green spaces goals from plans/policies Urban green spaces implementation Urban climate governance	⑦ **Outcome**
		25. Administrative accountability within own level	2	6	6	Urban green spaces goals from plans/policies Urban green spaces implementation Urban climate governance	⑦ **Outcome**
5. Conformance to general morality	Institutional	26. Information transparency	2	6	6	Urban green spaces goals from plans/policies Urban green spaces implementation	⑦ **Outcome**
		27. Legitimacy (conformity to the law)	2	6	7	Urban green spaces goals from plans/policies Urban green spaces implementation Formal changes Informal changes	⑦ **Outcome**

(continued)

Table 6.4 (continued)

First tier and structural criteria	Dimensions	Second tier criteria	First case	Second case	Third case	Corresponding subcodes (subthemes)	Corresponding categories
		28. Ethical acceptability	4	5	7	Urban green spaces goals from plans/policies Urban green spaces implementation Formal changes Informal changes	⑦ **Outcome**

1	Very clearly negative	7	Very clearly positive
2	Clearly negative	6	Clearly positive
3	Slightly negative	5	Slightly positive
N/A	Not sufficient information to assess	4	Neutral/insignificant

From the first to the third cases, **an increasing trend is identified in the following criteria**:

1. technical/ecological adaptability, resilience, and robustness,
2. inclusion of health and well-being benefits,
3. inclusion of rights to access information,
4. sustainability in land-use,
5. information transparency,
6. legitimacy (conformity to law),
7. ethical acceptability.

6.5.1 Summary

Results from the MCA show that **all three case studies are efficient in terms of planning and implementation within relatively short periods**, despite not all having to be successful, possibly due to the lack of information or understanding of the fundamental challenges of Urban green spaces planning within that particular time and social-economic development context—an information cost that the actors missed to consider. It also shows **high governance innovativeness in all three cases on each own terms, despite not all being successful**. In this first case, the innovativeness in governance is reflected by the fact that the Panyu Ecological Corridor was promoted as a conceptual plan and the weigh-ins of urban planning experts, which was considered innovative then. In the second case, the Haizhu Wetland Park project was the first to have a land property suitable transfer scheme for collective agricultural land conservation and prevention of urban sprawl. The second case (Haizhu Wetland Park) and the third case (Tianhe Sponge City demonstration site) score high.

Regarding adaptability, resilience, and robustness, there is an overall increasing trend from the first to the third case in the technical/ecological dimension of adaptability, resilience, and robustness. There is **also a generally growing trend for including health and well-being benefits and the rights to access information, legitimacy (conformity to the law), and ethical acceptability in all the case studies**. Compared to the first case, the Tianhe Sponge City project is backed by technically more proficient and meticulous plans and greater central government preference. Compared to the second case, the Tianhe Sponge City demonstration site construction was more straightforward and did not involve property rights complications or social justice issues of urban migrants.

The first case—the Panyu Ecological Corridor project, scores high on economic predictability and transferability of practices. However, these criteria do not necessarily guarantee success. The high economic predictability also indicates little intervention incurred, alias the "business as usual" path dependency. **The second case—the Haizhu Wetland Park transformation project, scores slightly lower on social equity and conformance to general morality**. Compared to the first case, the difference in the second case began with the institutionalisation of land property rights changes from collective ownership to state ownership. Besides, the second case also benefited from a much larger budget, which sorted out the direct cost–benefit equation better than in the first case for both local government and non-state actors. Furthermore, the second case was facilitated by a greater political preference from the central level and the emergence of the new national development philosophy *of Ecological Civilisation.* **The comparatively lower score in social equity results from the project´s inability to improve the living environment of the urban migrants inclusively within the wetland area due to the lack of institutional support for the urban migrants, which is probably too costly to combine with this project**.

References

Crawford SES, Ostrom E (1995) A grammar of institutions. Am Polit Sci Rev. https://doi.org/10.2307/2082975

Ostrom E (2005) Understanding institutional diversity. Princeton University Press

Chapter 7
Discussions

7.1 Summary and Discussion of Main Findings

7.1.1 Land Property Rights Arrangements Determine Urban Green Space Governance

In Guangzhou, land property rights, as well as actors who have the power to influence land property rights arrangements, determine the extent to which they participate in Urban green spaces governance, which also indirectly indicates that Urban green spaces resources in China are still governed primarily in the same way as land resources. Under this premise, the Guangzhou government plans and governs urban green spaces proactively by first obtaining land property rights, including ownership and management rights, and legalising those Urban green spaces plans under the guidance of urban planning experts. Through this approach, the municipal government can gain better control of the overall land-use in urban development. For example, in the second case, the municipal government acquired the collectively owned farmland and transferred it into a state-owned property while maintaining the agricultural function of the land. In the third case, the Guangzhou government first created a legally binding special plan before implementing the Sponge City Initiative based on a specific technical guideline produced by the central government.

From the CPR theory point of view, well-defined property rights help solve the problem of over-appropriation and free-riding of urban land resources. Clarifying property rights can enable individuals or collective actors in Urban green spaces governance to fully enjoy the goods or products they own and participate in Urban green spaces governance to a greater extent. This also explains why the participation of urban migrants in the Haizhu Wetland Park project (the second case) was relatively low, while the collective farmland owners could negotiate and bargain with the municipal and district governments. In this example, the environment of the urban

© Zhejiang University Press 2023

J. Liu, *Transforming Urban Green Space Governance in China Under Ecological Civilization: An Institutional Analysis*, Urban Health and Wellbeing,
https://doi.org/10.1007/978-981-99-6694-3_7

village where the migrants live has not been improved, which shows that the property rights transfer of the Haizhu Wetland Park project is not sufficiently inclusive, which externalises the cost of social equity. However, in this case, the ecological achievement is relatively successful because it redivided the physical boundaries of urban green space resources and the participation boundaries of the farmland owners and migrant workers. The farmland owners were able to negotiate with municipal governments and participate in decision-making, and for that, the Haizhu Wetland Park case can also be considered as having had a collective-choice arrangement to a certain extent. In addition, practices like the Panyu Ecological Corridor project (the first case) exemplified a simplistic Urban green spaces governance logic that treats Urban green spaces resources the same as the physical land. This perhaps overly simplistic arrangement was due to the low awareness of the actors on the ecosystem services that Urban green spaces provide or the lack of knowledge on the critical factors that influence the quality of Urban green spaces ecosystem services, including connectivity, biodiversity, and landscape diversity. The foregone benefits of Urban green spaces ecosystem services are essentially an opportunity cost.

Property rights arrangement is an essential formal institutional arrangement in CPR governance. Conceptualising Urban green spaces as CPR tells us that the property rights bundles for Urban green spaces governance should contain two separate sets—one for the urban land and the other for the Urban green spaces as urban ecological CPRs—each set with more refined definitions of the rights of access, withdrawal, management, exclusion, and alienation. As land resources in Chinese cities are usually scarce but not necessarily low exclusivity, land resources can be governed as CPR or private. Although the ecological functions and services of Urban green spaces are also scarce as urban land resources, they are shared by all citizens and are not exclusive. In this circumstance, properly defined rights of ownership, management, and withdrawal of urban green space resources could bring different outcomes. Based on this reasoning, a policy suggestion can be made to the Haizhu Wetland Park case. After obtaining the ownership of this agricultural land, it is entirely possible to allow the original farmers (or even enable the urban migrants) to continue managing their orchards and preserve the ecological functions as well as the local Lingnan ecological farming culture. The only property right that the Guangzhou Municipal Government needs to retain is to supervise the management and strengthen the standards and requirements in environmental conservation. This will mobilise non-state actors to participate in managing the Haizhu Wetland Park to a greater extent and reduce maintenance costs.

7.1.2 Guangzhou's Urban Green Spaces Governance Mostly in Line with Characteristics of Successful CPR Governance Regimes

Secondly, although China's Urban green spaces are mainly governed as land resources, their ecological functions, including the potential to cope with climate change, have gradually been formalised in the institutional arrangements in the past two decades. Guangzhou's Urban green spaces governance is mostly in line with the eight common characteristics of successful CPR governance regimes. Under the concept and national environmental governance framework *of Ecological Civilisation*, the focus on environment and climate change, the number and intensity of legally binding Urban green spaces targets in national policy documents increased, and that implies enhanced consideration of social costs and environmental benefits of Urban green spaces, which shows increasing clarity in the definition of CPR resource boundaries. Besides, the rigorous internal government reform and stricter environmental regulations and their enforcement show increasing clarity in the definition of CPR user boundaries, more arrangements for collective/choice or public participation, and improved monitoring, sanctions, and conflict resolution mechanisms. From the first to the third case, Guangzhou's Urban green spaces governance increasingly considered Urban green spaces resources as part of a larger and more complex social-ecological system and improved multi-level government supervision, reflecting the character of nested enterprise in successful CPR governance regimes. **However, not all the eight common characteristics of successful CPR governance regimes were identified in Guangzhou's Urban green spaces governance**. There was no minimal recognition (by state actors, e.g., upper-level governments) of the rights to organise as State actors dominate Guangzhou's Urban green spaces governance. By default, and need not be recognised by a higher-level government, the municipal government has both the right and (more as) responsibility to organise Urban green spaces governance together with non-state actors. This demonstrates the decisive role of China's institutional context in its CPR governance arrangements.

However, maintaining clear physical boundaries might not work the best for resources like urban green spaces as the way and physical extent to which ecosystem services benefit urban areas are fuzzy. For this reason, urban green space governance might be better governed with a fuzzy boundary which recognises the complexity of ecological compositions of urban green spaces. This again indicates the need to shift out of the current land property rights-based approach to urban green space governance. The idea that fuzzy resource boundaries could be more advantageous for promoting the ecosystem services of urban green spaces for climate adaptation, urban environmental health, and well-being contradicts the theory of CPR governance. But how to govern CPR with fuzzy physical boundaries? This is a real challenge to be researched and explored for solutions.

Except for the boundary clarity issue, which is now debatable, **governing Urban green spaces as CPRs in Chinese cities can be improved in the following areas: proportional equivalence between benefits and costs for all actors and broader**

collective-choice arrangements. To achieve proportional equality between benefits and costs for all actors, actors first need to gain a balanced and better understanding of Urban green spaces as an essential part of the urban complex social ecosystem and all the benefits and costs associated with Urban green spaces governance, including the ecological value Urban green spaces and the social value of collective choices made by non-state actors (the opposite of which are the costs of the ecosystem and social justice). A better understanding of ecological value is essential to reduce information acquisition costs gradually. Collective choice by non-state actors can help reduce the differences in the interest or willingness of actors in urban governance, thereby reducing the costs of cooperation.

7.1.3 Guangzhou's Urban Green Spaces Governance Attunes with Ecological Civilisation

Third, Guangzhou's Urban green spaces governance tends to attune with the national environmental governance framework *Ecological Civilisation* through conducting both the *means* and *ends* of institutional change. Guangzhou's Urban green spaces governance gradually went through these formal institutional changes from Case 1 to Case 3, focusing on changing the *means* of Urban green spaces institutions:

- the ecologicalisation in evaluating and promoting government officials' performance, for example, institutionalising mechanisms of self-check regarding environmental deeds
- the unification and systematisation of urban planning with reductions in the types of plans and the numbers of departments in charge, but with an increase in land-use restrictions
- increasing complexity in property rights arrangements
- government reform in critical aspects such as the inclusion of societal actors and the rules of the law.

All of these are institutional changes of the *means*. These four formal institutional change trends have been advanced through small-scale trial-and-error experiments, first at the local levels and, once proven effective, promoted by the central government and scaled up nationwide. From Case 1 to Case 3, Guangzhou's Urban green spaces governance went through several informal institutional changes which focused on changing the *ends* of Urban green spaces institutions, including enhanced notions of urban–rural coordination, public participation, ecological planning, environmental awareness among all actors, and citizens' role for government supervision.

Regarding value propositions, Case 1—the Panyu Ecological Corridor explicitly sought ecological adaptability on a conceptual plan but implicitly and predominantly pursued economic interests. Case 2—Haizhu Wetland Park was able to balance between seeking explicitly ecological values and compensating implicitly with economic development opportunities. Case 3—the Tianhe Sponge City demonstration site focused explicitly on ecological resilience, health benefits, and technical robustness. Oddly, even when these explicit ecological value propositions are evident in all three cases, they rarely directly address the issue of how these Urban green spaces resources might tackle climate change impacts. This is possibly due to:

- a gap in climate change perception between the central and local governments
- the technical-scientific challenge to confirm the direct causal relations between the local economic activities and regional climate change impacts, or to quantify the ecosystem services of Urban green spaces adequately
- that it is a strategy with differentiated focus distributed across levels of government and scales of time, population, and resources.

7.1.4 Sustained Barriers to Institutional Change for Urban Green Spaces Governance in China

Fourthly, three main institutional barriers/challenges remained in Guangzhou's urban green space governance despite substantial progress under Ecological Civilisation:

- **the lack of legal foundations for regular ecological status assessments**
- **low awareness of local state actors on climate change impacts and the ecological potential of Urban green spaces**
- **the lack of long-term commitment to governing urban green spaces based on recognising their ecological values and potential.**

First, Chinese cities have developed a path of dependence on land-based and economy-based urbanisation in the last few decades. A rigorous set of performance evaluation systems and governance trends based on economic development was established for local officials. In this context, the central government's political determination and recent ambition for *Ecological Civilisation* are plausible but not sufficient for urban green space governance to succeed. **This institutional barrier impedes multi-level institutional coordination, which requires solutions that establish the legal foundation for regular and long-term ecological status assessments so that Urban green spaces governance in Chinese cities can become more nature- and ecosystem-based in the long run.** Second, due to the current definition of Urban green spaces as a type of urban development land, urban green spaces in Chinese cities have been chiefly regarded as green infrastructure to serve urban economic development, which is a rather pragmatic and utilitarian approach that is no longer very suitable for the high complexity and uncertainty in the urban social-ecological system today. Under this premise, actors in urban green space governance fail to consider many ecological functions that are not yet entirely quantifiable in scientific research or monetisable in policy design. **Local state actors might not be**

as proactive in improving their awareness or collaborating with other departments on addressing climate change impacts through better urban green space governance. This institutional rigidity limits the horizontal interplay within local governments that requires institutional solutions. The third aspect relates to the dimension of time. Chinese city government officials are appointed and transferred hierarchically or from one city to another by the central government with a certain frequency to prevent corruption from emerging. Under this premise, local officials are required or are under political pressure to perform during their tenure. Hence, compared to urban green space governance projects that require a long-term commitment and are often less economically and politically impressive, most local officials prefer to build a visible and highly efficient infrastructure or economic project, e.g., host a G20 meeting or train station. **This institutional rigidity implies conflicting objectives, tensions, and trade-offs in political dimensions, which requires institutional solutions to foster long-term commitment to govern urban green spaces genuinely based on recognising their ecological values and potentials**.

7.2 Policy Implications

It is hardly straightforward to resolve these remaining challenges as they are institutional barriers that require fundamental change within the legal and administrative system of the country across levels as well as in the intrinsic values of all actors on a large scale. These changes require a high learning and communication capacity among all actors to reach a consensus on such institutional changes, build trust among each other, and support their efforts to achieve them. These institutional changes can be hugely costly in time, financing, political inconveniences, and social stability. Design more integrated urban green space plans that would also be more effective in addressing social issues requires reforms in the urban planning system, the interdepartmental coordination mechanism, the household registration system (Hukou) and, most importantly, change the narrow framing and targeting of urban green space planning in policy documents towards more comprehensive framing. To enable more public participation in planning and governance and not just supervision, it would also require an institutional change to enlarge the rights of citizens to create legal space for them. It also requires normative changes in how government officials and citizens perceive their relationships from the conventional "the governor" and "the governed" relationship to one that recognises all urban actors' capacity and willingness to contribute to improving the urban living environment as a *commons*. To ensure the institutional sustainability of these interventions, when one type of urban green space intervention is proven effective, there is decent institutional support to enable replicating in other parts of the city or upscaling in other cities. Finally, a set of intrinsic values that are more ecologically embedded and common-pool resource (CPR) oriented among all actors would also be effective and efficient for promoting more ambitious urban green space planning decisions.

The main findings presented above are expected to have the following implications on the Urban green spaces policies in Chinese cities for climate adaptation, urban health, and well-being.

1. Influence on the current practice of information public exposure, with the tendency to extend the time and scope of exposure so that citizens could be better informed about the respective Urban green spaces plans and be in a proper position to supervise local governments during the implementation process, thereby securing citizen's right to know and improving public acceptance.
2. Influence on the current speed of urban green space planning, with the tendency to allow more time in the planning process. This would enable more internal synergy to rise or increase between departments. In addition, it would allow more feedback to come from non-state actors, including urban migrants, which would make the Urban green spaces plans more socially inclusive, alias, and people-centred.
3. Influence on the institutional foundation of urban green space planning, with the tendency to create a more solid institutional foundation to improve the administrative accountability of respective departments and the rights and responsibilities of other actors.
4. Influence on the science-policy nexus, with the tendency to commission more scientific studies to enhance the scientific rigour, feasibility, and success rate of the Urban green spaces plans for climate change mitigation and adaptation.
5. Influence the way local governments narrate and understand climate change, with the tendency to enhance the narratives of climate change locally and make the nexus between Urban green spaces and climate change more pronounced to strengthen citizens' awareness of climate change and mobilise more local and voluntary climate actions. Finally, the logic that ecosystem services need to be more adequately quantified and monetised to facilitate more ambitious Urban green spaces planning decisions might need to be contested; instead, a set of intrinsic values that are more ecologically embedded and common-pool resource (CPR) oriented among all actors would also be effective and efficient for promoting more ambitious decisions on Urban green spaces planning.

Chapter 8
Conclusions

8.1 Research Overview and Reflection of Main Findings

Facing expeditious urbanisation and climate change impacts, how has China governed urban green spaces? This book has presumed the task of finding out. Urban green spaces are an entry point where actions and results matter for every urban citizen for better climate resilience, health, and well-being regardless of the social, economic, and political contexts in which they live. Due to the scales of its economy, population, and total carbon footprint, China bears a huge responsibility and potential and has seen solid political determination to act upon climate and environmental challenges amidst constant urbanisation and economic development needs. So how did Chinese cities approach the planning and governance of Urban green spaces? What institutional facilitation or challenges have they received or confronted when planning and managing Urban green spaces? How have local governments managed to upscale Urban green spaces towards more effective Nature-based Solutions? What are the good practices to be shared?

Furthermore, why have some challenges persisted despite the centralised government system and strong political determination? This book sets out to study three case studies on urban green space planning and governance in Guangzhou, a highly populated, compact, vulnerable coastal city in Southeast China. The aims were to understand the institutional dynamics, facilitators, and underlying barriers that could be at play in urban green space planning and governance, examine the extent and approaches for urban green space upscaling, analyse the costs and benefits taken into consideration, and understand the intrinsic-value-related factors of institutional barriers, which is essential for institutional innovation for promoting more Nature-based Solutions.

Based on insights from literature review of the Chinese and Western theories and practices of urban planning, and urban governance from the institutional perspective, this book builds a theoretical framework of common-pool resource (CPR) governance developed by American political scientist Elinor Ostrom to proceed. The theoretical

© Zhejiang University Press 2023

J. Liu, *Transforming Urban Green Space Governance in China Under Ecological Civilization: An Institutional Analysis*, Urban Health and Wellbeing,
https://doi.org/10.1007/978-981-99-6694-3_8

framework establishes Urban green spaces as an essential part of the complex urban social-ecological system that should theoretically be governed as a common-pool resource (CPR), with formal and informal institutional arrangements designating property rights and responsibilities to both state and non-state actors appropriate to respective levels and scales. The theoretical framework also expounds on how institutions evolve or change along with the changing transaction costs or the changing perceptions of them between actors and action situations. The breakdown of transactional cost–benefit balances constitutes a major driver of institutional change, formal or informal, leading to two types of institutional change: the means and ends. The book selects three empirical cases and uses the Institutional Analysis and Development (IAD) framework to structure a case study-based qualitative content analysis and an MCA, with data collected from semi-structured in-depth interviews and a selection of national and local Urban green spaces-related policies and plans.

The qualitative content analysis and an MCA yielded the following findings. First, land property rights and actors who have the power to influence land property rights arrangements determine the extent to which they participate in urban green space governance, which also indirectly indicates that urban green space resources in China are still governed primarily in the same way as land resources. Property rights arrangement is an essential formal institutional arrangement in CPR governance. Conceptualising urban green spaces as CPR tells us that the property rights bundles for Urban green spaces governance should contain two separate sets—one for the urban land and the other for the Urban green spaces as urban ecological CPRs—each set with more refined definitions of the rights of access, withdrawal, management, exclusion, and alienation. In addition, although China's Urban green spaces are mainly governed as land resources, their ecological functions, including the potential to cope with climate change, have gradually been developed into formal institutional arrangements in the past two decades. Guangzhou's Urban green spaces governance is mostly in line with the eight common characteristics of successful CPR governance regimes. Governing Urban green spaces as CPRs in Chinese cities can be improved by striking a better equivalence between benefits and costs for all actors and broadening the extent of collective-choice arrangements.

The above two findings indicate that Urban green spaces, as an essential part of the complex urban social/ecological system, cannot be simply governed as land resources. It is necessary to define them with an exclusive and clear set of property rights to attach importance to the ecosystem services and ecological values of Urban green spaces. In addition, over-exploitation and free riding of CPRs tell us that recognising human beings' irrational and rent-seeking nature is the prerequisite for successful human cooperation and sustainable CPR governance.

Furthermore, Guangzhou's Urban green spaces governance tends to attune with the national environmental governance framework *of Ecological Civilisation* through conducting both the *means* and *ends* of institutional change. Finally, despite substantial progress under *Ecological Civilisation*, three main institutional barriers/challenges remained in Guangzhou's Urban green spaces governance:

- the lack of legal foundations for regular ecological status assessments

- low awareness of local state actors on climate change impacts and the ecological potential of Urban green spaces
- the lack of long-term commitment to governing urban green spaces based on recognising its ecological values and potential.

The lack of legal foundations for regular ecological status assessments is an institutional barrier that impedes multi-level institutional coordination, which requires solutions that establish the legal foundation for regular and long-term ecological status assessments so that urban green spaces governance in Chinese cities can become more nature- and ecosystem-based in the long run. Local state actors might not be as proactive in improving their awareness or collaborating with other departments on addressing climate change impacts through better Urban green spaces governance. The low awareness of local state actors on climate change impacts and the ecological potential of Urban green spaces is an institutional rigidity that limits the horizontal interplay within local governments that requires institutional solutions. The lack of long-term commitment to governing urban green spaces based on recognising their ecological values and potentials is another institutional rigidity that implies conflicting objectives, tensions, and trade-offs in political dimensions, which requires institutional solutions to foster long-term commitment to govern Urban green spaces, recognising its ecological values and potentials.

These findings indicate that some institutions can achieve specific goals efficiently, but one cannot tell whether they can be sustained in the long run. The CPR theory tells us that institutional arrangements for long-term sustainable resource governance should provide individual and collective actors with the time and opportunity to participate in the process thoroughly and achieve the end goals/benefits. In Urban green spaces governance, a set of long-term, sustainable institutional arrangements should allow individual or group actors to participate in the governance process and fulfil their needs for good health, well-being, and a resilient environment facing climate change impacts.

8.2 Reflecting on the Aims, Objectives, and Methodology

This research has fulfilled the aims set forth at the beginning, some better than others. The theoretical framing and the empirical inquiry succeeded in capturing the institutional dynamics, facilitators, and underlying barriers in Guangzhou´s Urban green spaces governance. The empirical analysis also successfully examined the extent and approaches for upscaling Urban green spaces and improving Urban green spaces governance in Guangzhou. Costs and benefits taken into consideration by state- and non-state actors were not understood by the researcher as clearly as desired, partially because some costs and benefits related to ecosystem services and social equity could not be decently quantified or compared. Finally, some intrinsic-value-related factors of institutional barriers were identified, but more research is needed to understand how these intrinsic-value-related factors influence Guangzhou´s Urban green spaces governance, in which the researcher did not intend to invest. This research also has realised its objectives of fulfilling the research gap of institutional insights of Urban green spaces governance within the Chinese context, sharing successful practices, providing new insights from the institutional analysis based on the common-pool resource (CPR) theory for local state actors, and verifying the potential of system reform for achieving *Ecological Civilisation* through identifying existing institutional barriers and proposing solutions.

Regarding the methodology, this research has represented an earnest attempt to tackle the methodological challenges in measuring and evaluating institutional arrangements and institutional change by establishing comprehensive criteria, combining in-depth interviews with official policies and plans, and conducting an in-depth qualitative content analysis and a comprehensive MCA. It is also an attempt to approach the subject of sustainable development and climate change policies without getting too much subjected to normative or ideological influences. The researcher has attempted to create case study representativeness as high as possible by choosing those that are in the highest contestation of land-use tension, environmental conflicts, climate impacts, economic development needs, and urban–rural division, and by setting the study period for 20 years to allow pieces of evidence of institutional evolution or change to be included. The use of semi-structured in-depth interviews without time limits allowed a large amount of information to surface for this subject. The shortcoming of this semi-structured in-depth interview method is the substantial quantity and heterogeneity of the information generated, which complicated the analysis process.

Nevertheless, given that this is qualitative research, the large quantity and diversity of information can still be considered advantageous. Furthermore, the professional Qualitative Data Analysis Software MaxQDA has enabled a robust and thorough content analysis. Finally, a multi-criteria assessment (MCA) enabled a comprehensive evaluation of the three Urban green spaces planning and governance cases. The positive side of the MCA is its multiple references as the foundation, while the downside is that the MCA was conducted only by the researcher. The result might have been even more compelling if conducted both by the researcher and the interviewees.

8.3 Policy Recommendations

Based on the thorough institutional analysis of the local urban green space governance in Guangzhou, China, the researcher presents the following recommendations for policymakers for better urban green space governance practices in future:

- Recognise Urban green spaces as an essential part of the complex urban social/ecological system and attach importance to the ecosystem services and ecological values of Urban green spaces by defining them with an exclusive and clear set of property rights.

- Extend the time and scope of information public exposure for individual or group actors to be better informed and fully participate in Urban green spaces governance.
- Allow more time in the Urban green spaces planning process to enable more internal synergy to raise or increase between departments. This would also allow more feedback from non-state actors, including urban migrants, thereby improving inclusivity.
- Improve the accountability mechanism so that state- and non-state actors could better fulfil their rights and responsibilities.
- Commission more scientific research on the science-policy nexus of Urban green spaces to enhance the scientific rigour and policy effectiveness of Urban green spaces governance for climate change adaptation, urban health, and well-being improvement.
- Enhance the local narratives of climate change and the nexus between Urban green spaces and climate change to improve citizens' awareness of climate change.
- Accurate monetisation of ecosystem services could facilitate more ambitious Urban green spaces governance, yet it has remained a scientific challenge. Instead, local governments may focus on cultivating intrinsic values for citizens that treat Urban green spaces as CPR and are more ecologically oriented.
- Increase the number of long-term Urban green spaces governance arrangements and foster a more systemic approach to climate resilience, urban health, and well-being across administrative levels for the middle to long term.

8.4 Key Scientific Contributions and Limitations of This Research

8.4.1 Key Scientific Contributions

One important contribution of this book is the application of the common-pool resource (CPR) theory to understand CPR in both urban and Chinese socio-economic and political settings. This research complements existing studies on urban green space governance, which primarily depart from the perspectives of urban planning, landscape ecology, urban ecology, or political economy. The CPR theory has been well used in many countries such as Nepal, Switzerland, Brazil, India, and the USA to understand resource sustainability governance. However, it has been rarely applied to analyse empirical resource governance in Chinese socio-economic and political settings. The concepts such as property rights bundles, collective-choice arrangements, transaction costs, and institutional change yielded new insights into the power dynamics and interactions between actors, policy formation and dissemination processes, and the central-local government coordination and their evolution trends. These new insights have helped researchers understand why sometimes even ecologically conscient, socially just, and well-versed urban plans could fail.

Besides, this research tested the applicability of the CPR theory—a set of concepts and analytical elements created by scholars within the Western social and economic context to case studies in the Chinese urbanisation context, especially at the local level, which is broadly considered distinctive to the Western ones. There are limits to doing so. One of such limits might be the unfamiliarity of the interviewees, for which they often were surprised by the questions asked. Another limitation is that the CPR theory could not explain every finding in the empirical analysis. Nonetheless, these reactions precisely showed the novelty of the perspective introduced and the potential to yield undiscovered insights. However, at the same time, applying these so-called Western concepts and analytical elements in the Chinese context also helps to reduce the limits of the results only being applicable back to Western realities.

Another important scientific contribution of this research is identifying the need to define Urban green spaces with a set of property rights different from those of land resources. It is necessary to represent them with an exclusive and clear set of property rights to attach importance to the ecosystem services and ecological values of Urban green spaces,

Furthermore, this book has provided a detailed illustration to show how China's urban planning has primarily aligned with the evolutionary trends in Western urban planning theories. China's urban planning practices evolved, from:

1. viewing cities as simple, physical to complex, social-ecological systems
2. focusing on the comprehensive "end-state" of planning and the aesthetic-oriented "city beautification" movements to the rational process of planning
3. context insensitive (political, economic, and environmental) to context-sensitive,
4. local non-state actors exclusive to local actors inclusive
5. focusing on the planning method to the practical problems that planning seeks to resolve.

8.4.2 Limitations of the IAD Framework and the MCA Method for This Research

Although scholars are increasingly finding the IAD framework convincing for analysing a vast range of action situations thanks to the powerful meta-language that Ostrom (2005) developed after decades of theoretical inquiry and empirical testing, there are some limitations in using the IAD framework to analyse urban green space governance in the Chinese context. First, the IAD framework was developed by Ostrom (2005), an American political scientist who drew empirical evidence mainly from the US and other less centralised political contexts worldwide. Compared to most empirical evidence Ostrom had drawn on, the Chinese socio-economic and political contexts are drastically different in many ways: geographical and demographical sizes, the pace of environmental change, the tension between economic development and environmental protection, the systematic effect of policy change orchestrated by the central government, cultural perceptions of the environment, and access to official information by the public. This book elaborated a detailed chapter on contextualisation and case introduction to compensate for this shortage (see Chap. 5). In addition, even though the meta-language of the IAD framework is powerful for conducting institutional analysis for CPR governance, the researcher also finds the meta-language too dry to convey a normative, value-embedded message, e.g., on climate resilience or environmental sustainability in general, which is reasonably necessary for this research. Although the MCA is an emerging method well praised for its capability to navigate complex, multi-, and interdisciplinary research (Chambwera et al. 2015), the method is sometimes criticised for being subjective, especially if it is used in non-participatory research.

8.5 Future Research Directions

A few future research directions emerged from this book. One of the directions can be picked up from the intrinsic-value-related institutional barriers identified in this research, and the new research can focus on how these intrinsic-value-related factors influence urban green space governance. Another future research direction is understanding what kind of property rights bundles could be suitable for CPRs like urban green spaces to reflect the ecosystem services and ecological values of urban green spaces more appropriately. Besides, the multi-criteria assessment (MCA) is a modern research method for governance studies. Thanks to its flexible, adaptive nature, future research can and should experiment with diverse ways of conducting MCAs to determine their potential for informing urban environmental and climate governance. Finally, the institutional change currently taking place in China under *Ecological Civilisation* is a fascinating phenomenon with immense implications across sectors, levels, and populations, which requires more research to understand further its patterns, approaches, and long-lasting effects on Chinese society and its sustainability in this rapidly changing and drastically uncertain post-pandemic era under climate change.

References

Chambwera M, Heal G, Dubeux C, Hallegatte S, Leclerc L, Markandya A, McCarl BA, Mechler R, Neumann JE, Calvo E, Iglesias A, Navrud S, Kairiza T (2015) Economics of adaptation. In: Climate change 2014 impacts, adaptation and vulnerability: part a: global and sectoral aspects. https://doi.org/10.1017/CBO9781107415379.022

Ostrom E (2005) Understanding institutional diversity. Princeton University Press

Correction to: Transforming Urban Green Space Governance in China Under Ecological Civilization: An Institutional Analysis

Correction to:
J. Liu, *Transforming Urban Green Space Governance in China Under Ecological Civilization: An Institutional Analysis*, Urban Health and Wellbeing,
https://doi.org/10.1007/978-981-99-6694-3

The original version of this book was published with the author name Franz Gatzweiler which has now been excluded from this title. The book has been updated with the change.

The updated version of this book can be found at
https://doi.org/10.1007/978-981-99-6694-3

© Zhejiang University Press 2025

J. Liu, *Transforming Urban Green Space Governance in China Under Ecological Civilization: An Institutional Analysis*, Urban Health and Wellbeing,
https://doi.org/10.1007/978-981-99-6694-3_9

Annex

Form of Informed Consent for Interview

Background

This interview aims to gather data for the Ph.D. research project "Governance of Urban green spaces in China—Institutional Theories and Analysis with Case Studies from Guangzhou" of Jieling Liu, Ph.D. Candidate in the Interdisciplinary Program on Climate Change and Sustainable Development Policies at the Institute of Social Sciences—University of Lisbon (ICS-ULisboa), Portugal, and junior researcher at the "Urban Health and Wellbeing—A Systems Approach" (UHWB) program at the Institute of Urban Environment—Chinese Academy of Sciences (IUE-CAS). The ideal candidates of this interview include but are not limited to local government officials, urban planners, urban governance consultants, environmental and climate policy scholars, landowners, land appropriators, and residents. The researcher declares that the proposed questions of this interview will be non–sensitive topics only, and will be towards adults, non-vulnerable populations only.

Conditions to participate

You are invited to participate in the semi-structured, in-depth interviews voluntarily. Your answers will be kept confidential and used solely for this research. All the interview data will be collected, either in audio recording or hand-written notes, depending on your preference, and will then be stored and treated in a confidential, anonymous and codified way.

By signing this consent form, you agree to participate in this semi-structured, in-depth qualitative interview with the researcher, Jieling Liu, and permit her to use your answers as data/evidence for her PhD research project. The researcher will retain one copy of the consent form signed by both parties. You will be given a copy of the consent form as a record for what you have signed up.

© Zhejiang University Press 2023

J. Liu, *Transforming Urban Green Space Governance in China Under Ecological Civilization: An Institutional Analysis*, Urban Health and Wellbeing,
https://doi.org/10.1007/978-981-99-6694-3

The semi-structured, in-depth interview could last between 1 and 2 h, depending on the development of the conversation.

The researcher Jieling Liu appreciates your collaboration!

Consent to Participate in Research

I, ________________, voluntarily agree to participate in this interview for the research.

1. I understand that even if I agree to participate now, I can withdraw at any time during the interview or refuse to answer any question without any consequences of any kind.
2. I understand that I can withdraw permission to use data from my interview within two weeks after the interview, in which case the material will be deleted.
3. Before this interview, I have had the purpose and nature of the study explained to me in writing, and I have had the opportunity to ask questions about the study.
4. I understand that participation involves sharing the insights of my work and my opinions directly or indirectly related to the three chosen cases of urban green space planning and governance in Guangzhou.
5. I understand that I will not benefit directly or indirectly from participating in this research.
6. I understand that all information I provide for this study will be treated confidentially.
7. I understand that in any report on the results of this research, my identity will remain anonymous. This will be done by changing my name and disguising any details of my interview that may reveal my identity or the identity of people I speak about.
8. I understand that disguised extracts from my interview may be quoted in the interviewer/researcher's published work, including scientific papers, dissertation, blog posts, and more.
9. I understand that signed consent forms, original audio recordings, and transcripts of my interview (if applicable) can be obtained from an online storage service shared by the researcher for the period up until one year after her dissertation.
10. I understand that I am free to contact any of the people involved in the research to seek further clarification and information.

Please choose the following options that apply:
I agree that my interview will be recorded by

- Audio
- Video
- Handwriting notes only

Names, affiliations and contact details of the interviewee:
Name:
Affiliation and contact details:
Date:

Signature:

The researcher believes the participant is giving informed consent to participate in this interview.

Names, affiliations and contact details of the researcher:

Name: Jieling Liu

Affiliation and contact details: Institute of Social Sciences—University of Lisbon jielingliu@campus.ul.pt

Date:

Signature:

Semi-structured In-Depth Interview Outline

Part of the research data in this book is obtained through semi-structured in-depth interviews with officials from the Guangzhou Municipal Government, urban planning professionals, urban governance consultants, academic researchers, urban land appropriators (private enterprises, etc.), and community/residents. This outline contains initial questions for identifying the expertise of interviewees.

Table A.1 presents the main research questions of this research and the questions formulated accordingly for the semi-structured in-depth interviews.

本文的部分研究数据通过与广州市政府官员、规划专员、城市管理顾问专家、学术研究员、城市土地使用者 (私营企业等) 、社区居民进行深度访谈所得。下表列举了本论文的主要研究问题，以及在此基础上提出的用之与采访对象交流的半结构深入访谈问题。

Selected Semi-structured In-Depth Interview Excerpts (Original in Chinese)

Zhao: *"The common practice of the central government is to give a general task and some general and guiding indicators, for example, building a low-carbon city. How does the local government implement it? Every city needs to make its plans according to local conditions and development needs. Some cities will focus on the development of green buildings. Some cities may have a better climate, and they would focus on developing more urban green spaces. In contrast, others may focus on optimising the industrial structure, penalising enterprises with high pollution and low efficiency, or ordering them to conduct improvements, or ordering them to close. In this process, the city government plays a highly coordinated and presiding role and work with the lower-level governments to achieve the tasks or goals given by the central government."*

(Originally in Chinese: "中央政府的惯例做法是给出一个总任务和一些普遍性的，指导性的指标。比如说，建设低碳城市。地方政府怎么实施呢？每个城市

Table A.1 Semi-structured in-depth interview questions and corresponding variables in the IAD Framework

Main research questions and sub-questions	Semi-structured in-depth interview questions			Corresponding variables
	Questions oriented for experts in land resources and land economics (government officials, land economics scholars and urban governance consultants)	Questions oriented for experts in urban landscape ecology/green space planning (government officials, urban planning scholars and practitioners, and residents)	Questions oriented for experts in institutions and governance (government officials and scholars)	
Q1. What institutional facilitations or challenges have Chinese urban governments received or confronted when planning and managing Urban green spaces? Q1.1 What are the formal policy facilitations and challenges that Chinese urban governments have received or confronted when planning and managing Urban green spaces? Q1.2 What are the informal normative facilitations and challenges that Chinese urban governments have received or confronted when planning and managing Urban green spaces?	Please answer all questions within the contexts of China's urbanisation and the city of Guangzhou 1. The ownership rights of urban land resources in China are public, and urban green spaces is a category of urban development land. What are the advantages and disadvantages of this arrangement, in your opinion, and why? 2. The processes of leasing, use and gentrification of urban land resources are often dominated by pursuits of economic benefits. Are the social and environmental benefits included in consideration when it comes to these activities? If so, how are these dimensions measured? Please exemplify using the practices of Guangzhou 3. What are the changes and tendencies in urban land resource management in the last 20 years? 4. Have these tendencies influenced the way urban green spaces are planned and managed? If so, how?	5. Most Chinese cities have gone through the path from expansive to intensive urban development. During this process, what has changed in the formal principles of urban planning? 6. Is there a general goal or standard for the entire country for urban green space planning? What is that? For what does this goal/standard serve, and does it have any disadvantages? 7. During the New Millennium Year, China redefined its urbanisation vision and made long-term urban development plans. What were the main objectives of Guangzhou's urban planning at that time? How is it different from the current goal? 8. The urban land resources in Guangzhou have become increasingly scarce in the past two decades. How is this scarcity reflected in the work of the urban planning department? How was the planning and creation of urban green spaces affected?	9. In the past two decades, have there been any policies or regulations that are beneficial to urban green space planning and governance at the central and local levels? If so, what are they? 10. What are the changes you have envisioned or observed happening within the city-level government, following the ministerial reform under the environmental governance framework "Ecological Civilisation"? 11. What do you think the reform of the government system will bring about changes in urban green space planning and governance?	**Formal Rules-in-use**

(continued)

Table A.1 (continued)

Main research questions and sub-questions	Semi-structured in-depth interview questions			Corresponding variables
	Questions oriented for experts in land resources and land economics (government officials, land economics scholars and urban governance consultants)	Questions oriented for experts in urban landscape ecology/green space planning (government officials, urban planning scholars and practitioners, and residents)	Questions oriented for experts in institutions and governance (government officials and scholars)	
	12. In your opinion, what constitutes the economic, social, and environmental values of urban green spaces, respectively? 13. What roles do you think urban green spaces should play in urban development? Why? 14. From the perspective of urban land resources and land economics, what do you think should be considered in the planning and governance of urban green spaces?	15. From the perspective of urban development comprehensive planning, what do you think should be considered in the planning and governance of urban green spaces? 16. From the perspective of landscape ecology, what do you think should be considered in the planning and governance of urban green spaces? 17. In the process of rapid urbanisation, urban density increases, and the socio-economic system is becoming increasingly complex, and the rules for the use of urban land resources are increasingly demanding. How do you think this reality affects the planning of urban green spaces? 18. The economic upgrades in cities have driven residents to have higher standards for the quality of life, social well-being, and for the urban environment. How do you think this reality affects the planning of urban green spaces? 19. The impact of climate change, such as heatwaves, heavy rains, floods, and air pollution, has increasingly challenged urban infrastructural planning. How do you think this reality affects the planning of urban green spaces?	20. In the New Millennium Year, the Chinese central government proposed a path of "Scientific Development", whilst nowadays the trend is "People-Centred Development". Many scholars also consider that China's urbanisation has primarily been "tech-centred"—both technically and technologically. What is your opinion on this? 21. Are there any elements in the local traditional culture beneficial for urban green space planning and governance? Is there any conventional urbanisation thinking that is unfavourable for urban green space planning and governance? Please exemplify with the city of Guangzhou 22. From the perspective of institutions and governance, what do you think we should consider in the planning and governance of urban green spaces?	**Informal Rules-in-Use**

(continued)

Table A.1 (continued)

Main research questions and sub-questions	Semi-structured in-depth interview questions			Corresponding variables
	Questions oriented for experts in land resources and land economics (government officials, land economics scholars and urban governance consultants)	Questions oriented for experts in urban landscape ecology/green space planning (government officials, urban planning scholars and practitioners, and residents)	Questions oriented for experts in institutions and governance (government officials and scholars)	
Q2. How have local governments managed to upscale Urban green spaces towards more effective Nature-based Solutions? Q2.1 What has been the approach of the Guangzhou local governments for planning new Urban green spaces and managing them? Q2.2 Has the Urban green spaces in Guangzhou been effective as an Nature-based Solutions and enhanced local climate resilience and urban health and wellbeing?	• **Case 1: Ecological Corridor Conceptual Plan** • **Case 2: Haizhu Wetland Park (converted from an orchard)** • **Case 3: Tianhe Sponge City Demo Site**			
	23. In your opinion, is the existing state of urban green space in Guangzhou restricted by particular land-use history, topography, or socio-economic factors? 24. What measures have been undertaken to increase land for green infrastructures in the last 20 years in Guangzhou? 25. Proportionally speaking, how much land within the city consists of urban green spaces 20 years ago and now?	26. What were the urban environmental conditions, community characteristics, and the central and local regulations/guidance taken into consideration in planning and management of Case 1 / Case 2 / Case 3? 27. Could you please list the stakeholders involved in the planning, funding, implementation, and management of Case 1 / Case 2 / Case 3? 28. How did the municipal government organise the planning of urban green spaces with other stakeholders (interactions with sub-level governments, public and private stakeholders, and citizens) in Case 1/ Case 2/Case 3? Have any investment or incentive policies been introduced to strengthen the implementation of these projects? 29. Was Case 1/Case 2/Case 3 a cost–benefit effective project? Who were the primary beneficiaries of this project? Who didn't benefit from it?	34. What are the short- and long-term objectives and strategic approaches in establishing and managing urban green spaces in Guangzhou? 35. To what extent are these plans supported by commitments and budget allocations?	**Biophysical Conditions** **Community Attributes** **Rules-in-use** **Actors** **Action Situation** **Interaction** **Outcome** **Evaluative Criteria**

(continued)

Table A.1 (continued)

Main research questions and sub-questions	Semi-structured in-depth interview questions			Corresponding variables
	Questions oriented for experts in land resources and land economics (government officials, land economics scholars and urban governance consultants)	Questions oriented for experts in urban landscape ecology/green space planning (government officials, urban planning scholars and practitioners, and residents)	Questions oriented for experts in institutions and governance (government officials and scholars)	
		30. How did the implementation of Case 1/Case 2/Case 3 come out? To what extent does the project meet the needs of residents? Did it contribute to improving the urban green space in Guangzhou in general? Why? 31. Were there any monitoring and performance evaluation schemes applied in Case 1 / Case 2 / Case 3? If so, what criteria were used? 32. What do you think of the quality (size, distribution/accessibility, biodiversity, connectivity, etc.) of urban green spaces that currently exist in the city of Guangzhou? 33. Do you think nature-based solutions such as the Tianhe Sponge City demo site can effectively restore the degraded urban environments? Do you think the current layout and management schemes of urban green spaces in Guangzhou can effectively address climate change impacts such as heatwaves, flashfloods, and air pollution on the city? Why?		

(continued)

Table A.1 (continued)

Main research questions and sub-questions	Semi-structured in-depth interview questions			Corresponding variables
	Questions oriented for experts in land resources and land economics (government officials, land economics scholars and urban governance consultants)	Questions oriented for experts in urban landscape ecology/green space planning (government officials, urban planning scholars and practitioners, and residents)	Questions oriented for experts in institutions and governance (government officials and scholars)	
Q3. What are the good practices to be shared? Q3.1 What are the good practices in land-use from the three case studies? Q3.2 What are the good practices in Urban green spaces planning from these case studies? Q3.3 What are the good practices in urban governance in general from these three case studies?	36. Are there any recommendable experiences related to the arrangements of land-use types and land property rights from the three case studies above to be shared?	37. Are there any good urban green space planning practices to be shared from the three case studies above? Please explain from the aspects of cost management, planning methods, stakeholder participation and implementation	38. Is there any good urban management practices to be shared in Guangzhou's experiences? Please explain from the aspects of institutional innovation	**Outcome Evaluation**
Q4. Why some challenges persisted despite centralised government system and strong political determination? Q4.1 How do the strong political determination and centralised government system manifest themselves? Q4.2 What are the remaining challenges to be resolved? Q4.3 Why is it challenging to resolve these remaining challenges?	39. How do the strong political determination and centralised system in the Chinese central government manifest themselves in urban land resource management? What does the concept 'Ecological Civilisation' entail? 40. What are the land-use related challenges revealed in the three case studies above to be resolved? Why is it challenging to resolve them?	41. How do the strong political determination and centralised system in the Chinese central government manifest themselves in urban green space planning? What does the concept 'Ecological Civilisation' entail? 42. What are the urban green space planning related challenges revealed in the three case studies above to be resolved? Why is it challenging to resolve them?	43. How do the strong political determination and centralised system in the Chinese central government manifest themselves in urban governance and institutions? What does the concept 'Ecological Civilisation' entail? 44. What are the urban governance and institutions related challenges revealed in the three case studies above to be resolved? Why is it challenging to resolve them?	**Outcome Evaluation**

都需要因地制宜，做出自己的方案。有些城市会专注于发展绿色建筑。有一些城市可能气候比较优良，会专注于城市绿化，还有一些城市会专注于优化产业结构，把高污染低效能的工业企业处罚或勒令改进，或勒令关门。这个过程中城市政府需要扮演一个高度协调主持工作的角色，与下级政府共同达成中央城府给出的任务或目标。”).

Xiao: *"For example, for constructing Low-Carbon cities, the mayor would discuss plans with city government officials and experts before making the final collective decision together with the Urban and Rural Planning Management Committee. If the mayor alone says that this plan is good, it is not possible (to pass). For example, the mayor said that he wanted to focus on building more urban green spaces, a department chief on land resources or a land economy expert would likely say that the cost of land resources for planting trees is relatively high and would ask the land economy experts to do some calculations. Planting trees may result in the loss of some more direct and more efficient economic benefits, such as building real estate. In such circumstance, the department chief of urban and rural construction may say that they happen to have a residential real estate development project, which could contribute to building Low-Carbon cities by building apartments that fit green building standards."*

(Originally in Chinese: 比如低碳城市建设，市长会与市政府委员以及行业专家一起讨论方案。最后城乡规划管理委员会一起来集体决定。如果市长他一个人说这个方案好，那是不可以的。比如这个市长说他想要专注于城市绿化，土地资源的政府委员或者专家很可能会说种树的土地资源成本比较大。让土地经济学专家算一下，可能种树会导致失去一些更高效率的经济收益，比如说盖房地产。这时候城乡建设的政府委员可能会说，我们正好有个开发商，要不让他建几万平米的房子全部让他用绿色建筑吧。”).

He: *"The Sponge City project has received strong support from the Guangzhou Municipal Government, and now they are establishing a demonstration site in Tianhe District. However, at the same time, the urban planning and construction departments are also experiencing resistance from the bureau of water resources. As the construction of Sponge City necessitates ecological and natural methods for flood retention and wastewater treatment, there will be fewer budgets in the future for the bureau of water resources and negatively impact its downstream stakeholders. Therefore, a conflict of interest has arisen between different departments within the municipal government, and this conflict can be understood as the coordination or negotiation cost within the municipal government."*

(Originally in Chinese: 海绵城市这个方案得到了广州市政府的大力支持，然后现在他们正在天河区建立一个实验点。但是同时，规划部门也在经受水利部门的一个阻力，因为建海绵城市就是用生态和自然的方式去排水，就意味着水利部门的预算会下降，其下游的水利工程利益相关者也会受到了影响。因此在市政府内部的不同部门间产生了利益冲突，这一冲突也就可以理解为政府内部的协调成本。”).

Yuan: *"Ordinary citizens have the legal rights to be informed and make decisions about the environment. The government generally has a process for publicizing information on construction projects to obtain public feedback on environmental impacts. In addition, the public can participate in the improvement of a particular project* *through supervision, and reporting, if they find certain decisions or actions of a government department or an official suspicious or improper."*

(Originally in Chinese: "公众有环境知情权和环境决策权，政府公开建设项目的信息一般有一个公示的过程以求得市民意见反馈。此外公众对于政府官员或部门有操作不当的嫌疑可以通过监督和举报的形式参与到项目的改善上来。")

Cai: *"In 2011, the Guangzhou Municipal Government submitted to the Ministry of Land Resources the proposal to acquire the Wanmu Orchard farmland without changing the land function. The Haizhu District Government then established a land acquisition work unit with an advocacy team and land acquisition and demolition team. The advocacy team informed the farmland owners with materials, including posters and videos of the farmland acquisition plan, through meetings with question-and-answer sessions with representatives."*

(Originally in Chinese: "广州市政府于2011年向国家国土部提交了只征不转的海珠万亩果园管理方案。海珠区政府随后成立了征地工作队，内设宣传组和征地拆迁组。宣传组所用的材料包括征地方案海报和视频，通过与村民代表开会和问答环节来普及方案内容。")

Cai: *"The representatives of farmland owners, whose role is to express opinions and make decisions for the collective benefits of the farmland owners, have a significant amount of power and long-rooted vested interests in the local social and economic network. The institutional arrangements to trade the farmland ownership for building the Haizhu Wetland Park are against their existing vested interests. Therefore, some farmland owner representatives were not enthusiastic about the proposal. They further negotiated with the municipal government on the financial compensation of farmland property right transfer."*

(Originally in Chinese: "村民代表作为村民民意的反映者以及利益安排的执权者，在当地的社会经济关系网络常年积累了诸多既得利益，并拥有大量权力。把农用地的所有权出让并改成海珠湿地公园的方案和制度安排对其原来的利益安排有所冲突，因此有部分村民代表的积极性并不高，进一步与市政府进行了出让农用地的价格的协商。"

Huang: *"The institutional arrangements for the Wanmu Orchard farmland acquisition ignored the impact on some actual residents of urban migrants in surrounding urban villages, intentionally or unintentionally. Transforming the orchard into the Haizhu Wetland Park is an improvement in the quality of the ecological environment in the area. It is clearly stipulated that 10% of the reserved land would be used for commercial development (hotels, shops,* etc.*) for the farmland owners with Guangzhou urban Hukou. These two interventions jointly increased the semi-informal housing prices in the urban villages. The leasing price also indirectly makes these urban villages more crowded, and the quality of life for those urban migrants who live there further declined. In this process, the migrant workers were indirectly negatively affected by the institutional arrangements for the Haizhu Wetland Park, and they were not officially notified, let alone new institutional arrangements to compensate for the negative impacts."*

(Originally in Chinese: "万亩果园征地的制度安排有意或无意地忽略了对周边村社中的部分实际居住者外来民工的影响。万亩果园升级为海珠湿地公园

是该区域的生态环境质量的提高, 明确规定10%留用地给拥有广州户口的村民作商业发展 (酒店、商铺等) 用途, 两者共同推高了村社内住房租赁的价格, 另外间接使得城中村更加拥挤, 生活质量进一步下降。这一过程中, 外来民工受到了万亩果园征地的制度安排的间接负面影响, 并未得到正式的通知, 更别论新的制度安排来补偿其所受到的负面影响。"

List of Interviewees

Number	Interview theme	Name	Profession	Affiliated Institution	Field of Expertise	Location of Interview	Type of Interview	Time of Interview
2	Land Economy and Institutions	***Ho**	Academic; governance consultant	Zhejiang University, Land Academy for National Development (LAND)	Land economy and institutions, resource and environmental management, land allocation efficiency and governance efficiency	Beijing	In person	12-Jul-18
3	Land Economy and Institutions	*** Tan**	Academic; governance consultant	Zhejiang University, Land Academy for National Development (LAND)	Land economy and institutions, resource and environmental management, land allocation efficiency and governance efficiency	Hangzhou	In person	4-Sep-18
4	Land Economy and Institutions	*** Yue**	Academic; governance consultant	Zhejiang University, Land Academy for National Development (LAND)	Urbanisation, land planning and spatial management, and big data	Hangzhou	In person	4-Sep-18
5	Land Economy and Institutions	*** Zhang**	Academic; governance consultant	Zhejiang University	Urban development and land institutions, development and governance of characteristic towns, urban environmental governance and sustainable development	Hangzhou	In person	6-Sep-18

(continued)

(continued)

Number	Interview theme	Name	Profession	Affiliated Institution	Field of Expertise	Location of Interview	Type of Interview	Time of Interview
6	Land Economy and Institutions	*** Gao**	Academic; governance consultant	Zhejiang University	Government management and innovation, local government public administration	Hangzhou	In person	6-Sep-18
7	Land and urban planning	*** Lin**	Urban planner	Guangzhou Land Resource and Planning Committee	Land planning, urban and rural planning, planning system	Guangzhou	phone call	24-Jan-18
8	Land Economy and Institutions	*** Wu**	Academic; governance consultant	Zhejiang University	Government management and innovation, public administration	Hangzhou	In person	1-Nov-19
9	Urban Planning, Environment and Development	*** Zhao**	Academic; governance consultant	Peking University	Urban planning, environment and development	Beijing	In person	13-Jul-18
10	Urban Planning	*** Lv**	Urban planner; local government official	Bureau of Urban Planning of Guangzhou	Land planning, urban and rural planning, preparation and approval of urban plans	Guangzhou	phone call	15-Aug-19
11	Urban Planning	***Rao**	Urban planner; local government official	Bureau of Urban Planning of Guangzhou	Land planning, urban and rural planning, preparation and approval of urban plans	Guangzhou	Social network message exchange	14-Aug-19
12	Urban Planning	*** Deng**	Urban planner; local government official	Guangzhou Urban Planning & Design Survey Research Institute	Land planning, urban and rural planning, preparation and approval of urban plans	Guangzhou	phone call	13-Aug-19
13	Urban Planning	*** Xu**	Urban planner; local government official	Guangzhou Urban Planning & Design Survey Research Institute	Land planning, urban and rural planning, preparation and approval of urban plans	Guangzhou	In person	14-Dec-17

(continued)

(continued)

Number	Interview theme	Name	Profession	Affiliated Institution	Field of Expertise	Location of Interview	Type of Interview	Time of Interview
14	Urban Planning	*** Liu**	Urban planner	Guangzhou Urban Planning & Design Survey Research Institute	Land planning, urban and rural planning, preparation and approval of urban plans	Guangzhou	In person	3-Jul-18
15	Urban Planning, Environment and Development	*** Yuan**	Academic, governance consultant, urban planner	South China University of Technology, Urban Planning Comittee of Guangzhou	Urban planning theory and design, regional and urban development strategy planning, urban master planning, urban renewal and land institutions	Guangzhou	In person	11-Sep-18
16	Urban Planning, Environment and Development	*** Ma**	Academic, governance consultant, urban Planner, local government official	Guangdong Urban and Rural Academy of Planning and Design, Urban Planning Comittee of Guangzhou	Urban planning theory and design, regional and urban development strategy planning, urban master planning, urban renewal and land institutions	Guangzhou	In person	28-Aug-18
17	Urban Planning, Environment and Development, Institutions of Planning	*** Zhao**	Urban planner, local government official	China Eco-City Academy	Urban planning theory and design, regional and urban development strategy planning, urban master planning, urban renewal and land institutions	/	phone call	29-Aug-18
18	Urban Planning, Environment and Development	*** A**	Academic, governance consultant, urban planner	Alexandria University Egypt	Urban planning theory and design, environment and development	Beirut	In person	28-Oct-18

(continued)

(continued)

Number	Interview theme	Name	Profession	Affiliated Institution	Field of Expertise	Location of Interview	Type of Interview	Time of Interview
19	Urban Planning, Environment and Development, Institutions of Planning	***Leung**	Academic, governance consultant	Queen's University	Philosophy and theory of urbanisation, urban planning theory and design, environment and development	/	via Email	15-Aug-19
20	Urban Planning, Environment and Development, Institutions of Planning	***Wu**	Academic, governance consultant	Bartlett School of Planning, University College London	Urbanisation, government management and innovation, local government public administration and institutions of land economics	Manchester	In person	1-Nov-18
21	Urban Planning, Environment and Development, Institutions of Planning, Urban Governance	*** You**	Governance consultant	Guangzhou Institute for Urban Innovation	Urban governance, urban government innovation, green development and urban intergovernmental collaboration	/	via Skype	2-Nov-17
22	Urban Planning, Environment and Development, Institutions of Planning, Urban Governance	*** Jon**	Academic, governance consultant	TU Delft Netherlands	Eco Cities, Place Branding, Corporate Social Responsibility, Inclusive Prosperity	Suzhou	In person	27-Jul-18
23	Urban Planning, Environment and Development	*** Yu**	Academic, governance consultant, urban planner	Peking University	Theoretical research and practice of urban and regional planning, urban and landscape design and ecological planning	Beijing	In person	13-Jul-18

(continued)

(continued)

Number	Interview theme	Name	Profession	Affiliated Institution	Field of Expertise	Location of Interview	Type of Interview	Time of Interview
24	Urban Planning	*** He**	Urban planner	Turenscape Urban Planning Consultancy	Urban and regional planning, urban and landscape design, ecological planning	Beijing	In person	13-Jul-18
25	Urban Planning, Environment and Development	*** Rong**	Academic, governance consultant	School of Architecture - Tsinghua University	Theory and method of ecological planning, green building, and urban ecological restoration	Beijing	In person	17-Jul-18
26	Urban Planning, Environment and Development, Institutions of Planning	*** Xiao**	Academic, governance consultant, urban planner	Guangdong Provincial Academy of Environmental Science	Theory and method of ecological planning, green building, and urban ecological restoration	Guangzhou	In person	31-Aug-18
27	Urban Planning, Environment and Development, Institutions of Planning	*** Li**	Urban planner	Guangzhou Urban Planning & Design Survey Research Institution, Urban Planning Research Center	Urban development strategy, low-carbon eco-city, historical and cultural city protection and urban renewal	Guangzhou	In person	15-Aug-19
28	Urban Planning, Environment and Development, Administrative Management	*** Cai**	Local government official	Management Office of the Haizhu Wetland Park, Haizhu District, Guangzhou	Urban planning, environment and development, administration	Guangzhou	In person	13-Aug-19
29	Urban Planning	*** Long**	Urban planner	Guangzhou Urban Planning & Design Survey Research Institute; Urban Planning Research Center / Eco-spatial Planning & Eco-Restoration	Urban development strategy, low-carbon eco-city, historical and cultural city protection and urban renewal	Guangzhou	In person	14-Aug-19

(continued)

(continued)

Number	Interview theme	Name	Profession	Affiliated Institution	Field of Expertise	Location of Interview	Type of Interview	Time of Interview
30	Urban Planning, Environment and Development, Administrative Management	***Huang**	Local government official	Bureau of Land-Use and Urban Planning of Haizhu District	Urban planning, environment and development, administration	Guangzhou	In person	13-Aug-19
31	Environment and Development, Administrative Management	***Zhang**	Local government official	Bureau of Forestry and Landscaping of Guangzhou	Administration and management of parks and other scenic areas	Guangzhou	phone call	11-Aug-19
32	Climate Change, Urban Health	***Huang**	Academic, governance consultant	School of Public Health, Sun Yat-sen University	Health Policy and Management, Climate Change and Health, Environmental Epidemiology	Xiamen	In person	4-Nov-19
33	Climate Change, Urban Health	*** Liu**	Academic, governance consultant	Chinese Center for Disease Control and Prevention	Health Policy and Management, Climate Change and Health, Environmental Epidemiology	Lisbon	In person	23-Sep-18
34	Climate Change, Urban Health	*** Zhu**	Academic, governance consultant	Institute of Urban Environment, Chinese Academy of Sciences	Environmental soil science and environmental biology	Xiamen	In person	27-Nov-17
35	Climate Change, Urban Health, Urban Governance	*** Mur**	Academic, governance consultant	Public Health England	Climate change, urban health, multi-agency collaboration, urban governance	Lisbon	In person	23-Sep-18

(continued)

(continued)

Number	Interview theme	Name	Profession	Affiliated Institution	Field of Expertise	Location of Interview	Type of Interview	Time of Interview
36	Climate Change, Urban Health, Urban Governance	***Ban**	Academic, governance consultant	Asian Development Bank	Health system governance, intersectoral and multi/stakeholder collaborative planning, strategic financial and institutional planning	Xiamen	In person	4-Nov-19
37	Climate Change, Urban Health, Urban Governance	*** Dor**	Academic, governance consultant	World Health Organization	Public health, global environmental health governance and justice	/	via email	Jun-20
38	Climate Change, Environment and Development, Intergovernmental Collaboration	*** Liu**	Governance consultant	C40 Climate Leadership Network	Climate change, urban health, multi-agency collaboration, urban governance	/	via email	3-Aug-18
39	Climate Change, Environment and Development, Intergovernmental Collaboration	*** Xie**	Governance consultant	C40 Climate Leadership Network	Climate change, urban health, multi-agency collaboration, urban governance	/	via email	3-Aug-18
40	Climate Change, Urban Governance	***Zhang**	Local government official	Guangdong Provincial Academy of Environmental Science	Regional atmospheric environment planning, policy, standards, and technology, water environmental management policy and mechanism	Guangzhou	In person	6-Aug-18
41	Climate Change, Environment and Development, Institutional Reform, Urban Governance, Ecological Civilisation	***Zhuang**	Academic, governance consultant	Chinese Academy of Social Sciences, CASS Thinktank for Eco-Civilisation Studies	Climate change and low-carbon economics, policy, development strategy and planning, and low-carbon city evaluation methods	Beijing	In person	12-Jul-18

(continued)

(continued)

Number	Interview theme	Name	Profession	Affiliated Institution	Field of Expertise	Location of Interview	Type of Interview	Time of Interview
42	Climate Change, Environment and Development, Institutional Reform, Urban Governance, Ecological Civilisation	*** Su**	Academic, governance consultant	Institutes of Science and Development - Chinese Academy of Sciences	Climate change policies, urban development paradigm, Ecological Civilization	/	via Email	22-Jun-18
43	Climate Change, Environment and Development, Institutional Reform, Urban Governance, Ecological Civilisation	*** Li**	Academic, governance consultant	Institute of Urban Development and Environment - Chinese Academy of Social Sciences	Climate change policies, urban development paradigm, Ecological Civilisation	Beijing	In person	13-Jul-18
44	Climate Change, Environment and Development, Institutional Reform, Urban Governance, Ecological Civilisation	*** Kang**	Academic, governance consultant	Development Research Center of Guangdong	Strategic issues of ecological construction and green development in Guangdong province, policies and regulations for environmental protection, energy conservation and emission reduction, climate change adaptation, and institutional reform under Ecological Civilisation	Guangzhou	In person	6-Aug-18
45	Climate Change, Environment and Development, Institutional Reform, Urban Governance, Ecological Civilisation	*** Luo**	Academic, governance consultant	Institute of Urban Development and Environment - Chinese Academy of Social Sciences	Environmental economics, urban economy and sustainable development, eco-city construction	Beijing	In person	14-Jul-18

(continued)

(continued)

Number	Interview theme	Name	Profession	Affiliated Institution	Field of Expertise	Location of Interview	Type of Interview	Time of Interview
46	Urban planning, inclusive development	*** Liu**	Urban planner, governance consultant	Morrow Architects & Planners Pte.Ltd.	Urban planning, landscape design, sustainable urban development, people-centred urbanism	/	phone call	29-Dec-17
47	Environmental information disclosure and public participation	*** Ma**	Head of environmental non-governmental organization	China Institute of Public & Environmental Affairs (IPE)	Water pollution prevention and control and environmental information disclosure, public participation	/	phone call	10-Jul-18
48	Urban Planning, Environment and Development, Institutions of Planning	*** Yao**	Urban planner, local government official	Guangzhou Urban Planning Research Center	Formulation and implementation of urban plans and regulations	Guangzhou	In person	15-Aug-19
49	Urban Planning, Environment and Development, Institutions of Planning	*** Wang**	Urban planner, local government official	Guangzhou Urban Planning Research Center	Formulation and implementation of urban plans and regulations	Guangzhou	In person	In person
50	Urban Planning, Environment and Development, Institutions of Planning	*** Zhang**	Academic, governance consultant, urban planner	Urban Planning Society of China	History of urban development and construction in China, the history of western urban planning thought, institutional reform and urban structuring in China, and the evolution of urban system planning and development in China	Beijing	In person	15-Jul-18

List of Policy Documents

Level	Theme	Title in Chinese	Title in English	Time of publication	Target period	Type	Publication entity	Coordination entity	Overarching political philosophy/policy principle
National	Climate change/ sustainable development	2001年中国国土绿化状况公报	Communiqué on the Land Greening Status in China 2001	2001	/	communique	National Greening Committee Office	/	Scientific Development & Harmonious Society
		2005 年中国国土绿化状况公报	Communiqué on the Land Greening Status in China 2005	2005	/	communique	National Greening Committee Office	/	Scientific Development & Harmonious Society
		国务院关于落实科学发展观加强环境保护的决定	Decision of the State Council on Implementing Scientific Outlook on Development and Strengthening Environmental Protection	2005	2010	strategic plan	State Council	All ministries and levels of government institutions	Scientific Development & Harmonious Society
		中国应对气候变化国家方案	China's National Program on Climate Change	2007	2010	strategic plan	State Council	All ministries and levels of government institutions	Scientific Development, Resource Saving & Environmental Protection
		国家环境与健康行动计划 2007-2015年	National Action Plan on Environment and Health (2007–2015)	2007	2015	strategic plan	State Council	National Environmental Protection Bureau & Healthy Ministry	Scientific Development, Harmonious Society, & People-Centred

(continued)

(continued)

Level	Theme	Title in Chinese	Title in English	Time of publication	Target period	Type	Publication entity	Coordination entity	Overarching political philosophy/policy principle
		中国应对气候变化的政策与行动	China's Policies and Actions on Climate Change	2008	/	report	State Council	/	/
		中国生物多样性保护战略与行动计划 2011—2030 年	China National Biodiversity Conservation Strategy and Action Plan (2011–2030)	2011	2030	strategic plan, action plan	State Council	National Environmental Protection Bureau	Scientific Development
		2012 年中国国土绿化状况公报	Communiqué on the Land Greening Status in China 2012	2012	/	communique	National Greening Committee Office	/	Scientific Development & Ecological Civilisation, & People-Centred
		国家大气污染防治行动计划	National Air Pollution Prevention and Control Action Plan	2013	2017	action plan	State Council	National Environmental Protection Bureau & Meteorological Administration	Scientific Development & Ecological Civilisation
		国家适应气候变化战略	National Strategy for Climate Change Adaptation	2013	2020	strategic plan	State Council	National & local departments of development and reform	Scientific Development & Ecological Civilisation, & People-Centred

(continued)

(continued)

Level	Theme	Title in Chinese	Title in English	Time of publication	Target period	Type	Publication entity	Coordination entity	Overarching political philosophy/policy principle
		国家应对气候变化规划**2014-2020**年	National Plan on Climate Change 2014–2020	2014	2020	action plan	National Development and Reform Commission	All government departments at city level	Scientific Development, Resource Saving & Environmental Protection, Ecological Civilisation
		环境保护法修订版	Revision of the Environmental Protection Law	2014	/	regulation	State Council	Environmental protection departments at national and local levels	Sustainable Development & Ecological Civilisation
		中国应对气候变化的政策与行动	China's Policies and Actions on Climate Change	2015	/	report	National Development and Reform Commission	/	/
		城市适应气候变化行动方案	Action Plan for Urban Climate Adaptation	2016	2020	action plan	State Council	All government departments at city level	Sustainable Development & Ecological Civilisation
		中国落实2030年可持续发展议程国别方案	China's National Plan on Implementation of the 2030 Agenda for Sustainable Development	2016	2030	strategic plan	State Council	All ministries and levels of government institutions	Innovation, Coordination, Green, Open, Sharing, Sustainable Development & Ecological Civilisation

(continued)

(continued)

Level	Theme	Title in Chinese	Title in English	Time of publication	Target period	Type	Publication entity	Coordination entity	Overarching political philosophy/policy principle
		中国应对气候变化的政策与行动	China's Policies and Actions on Climate Change	2018	/	report	Ministry of Ecology and Environment	/	/
		中国应对气候变化的政策与行动	China's Policies and Actions on Climate Change	2019	/	report	Ministry of Ecology and Environment	/	/
		2019 年中国国土绿化状况公报	Communiqué on the Land Greening Status in China 2019	2019	/	communique	National Greening Committee Office	/	"Two Mountains" theory & Ecologicl Civilisation
		"健康中国2030"规划	Healthy China 2030	2016	2030	strategic plan	State Council	All ministries and levels of government institutions	Sustainable Development
	Urban spatial planning and development	全国城市生态保护与建设规划 2015-2020年	National Plan of Urban Ecological Protection and Construction (2015–2020)	2015	2020	strategic plan	Ministry of Housing and Urban–Rural Development & Ministry of Environmental Protection	All ministries and levels of government institutions	Innovation, Coordination, Green, Open, Sharing
		国家新型城镇化规划 2014-2020年	National New Urbanization Plan (2014–2020)	2014	2020	strategic plan	State Council	All ministries and levels of government institutions	Scientific Development & People-Centred

(continued)

(continued)

Level	Theme	Title in Chinese	Title in English	Time of publication	Target period	Type	Publication entity	Coordination entity	Overarching political philosophy/policy principle
	Institutional reform	关于全面深化改革若干重大问题的决定	Decision of the Central Committee of the Communist Party of China on Some Major Issues Concerning Comprehensively Deepening the Reform	2013	2020	strategic plan	State Council	State Council	Scientific Development
		关于加快推进生态文明建设的意见	Opinions on Accelerating the Construction of Ecological Civilization	2015	2020	conceptual plan	State Council	All ministries and levels of government institutions	Scientific Development & Ecological Civilisation, & People-Centred
		生态文明体制改革总体方案	Master Plan of Ecological Civilization System Reform	2015	/	action plan	State Council	Central Leading Group for Deepening Overall Reforms & Special Group on Reform of the Economic and Ecological Civilization System	Scientific Development, Resource Saving & Environmental Protection, Human-Nature Harmony
		中共中央关于深化党和国家机构改革的决定	Decision of the CPC Central Committee on Deepening the Reform of the Party and State Institutions	2018	/	strategic plan	State Council	All ministries and levels of government institutions	Scientific Development

(continued)

(continued)

Level	Theme	Title in Chinese	Title in English	Time of publication	Target period	Type	Publication entity	Coordination entity	Overarching political philosophy/policy principle
		深化党和国家机构改革方案	Plan to Deepen Reforms in Party and State Institutions	2018	2019	action plan	State Council	All ministries and levels of government institutions	Scientific Development & People-Centred
	Macro socio-economic planning	中国国民经济和社会发展第十个五年计划	10th Five-Year National Socio-Economic Development Plan	2001	2005	macro socio-economic development plan	National Development and Reform Commission	All ministries and levels of government institutions	Development as Priority, People-Centred, Economy and Society Coordinated Development
		中国国民经济和社会发展第十一个五年计划	11th Five-Year National Socio-Economic Development Plan	2006	2010	macro socio-economic development plan	National Development and Reform Commission	All ministries and levels of government institutions	Scientific Development & People-Centred
		中国国民经济和社会发展第十二个五年计划	12th Five-Year National Socio-Economic Development Plan	2011	2015	macro socio-economic development plan	National Development and Reform Commission	All ministries and levels of government institutions	Scientific Development, Resource Saving & Environmental Protection
		中国国民经济和社会发展第十三个五年计划	13th Five-Year National Socio-Economic Development Plan	2016	2020	macro socio-economic development plan	National Development and Reform Commission	All ministries and levels of government institutions	Scientific Development, Innovation, Coordination, Green, Open, Sharing, People-Centred

(continued)

(continued)

Level	Theme	Title in Chinese	Title in English	Time of publication	Target period	Type	Publication entity	Coordination entity	Overarching political philosophy/policy principle
Municipal level	Climate change/ sustainable development	广州市城市总体规划 2001–2010	Master Plan of Guangzhou City 2001–2010	2005	2010	action plan	Municipal Government of Guangzhou	All ministries and levels of government institutions	Scientific Development, Sustainable Development & Harmonious Society
		广州市土地利用总体规划2006-2020	Land-Use Master Plan of Guangzhou City 2006–2020	2006	2020	action plan	Municipal Government of Guangzhou	Guangzhou Land Resources and Housing Administration	Scientific Development, Resource-Saving & Harmonious Society
		广州市城市绿地系统规划	Urban Green Space System Plan of Guangzhou City	2010	2020	action plan	Municipal Bureau of Parks and Forestry of Guangzhou	Municipal Bureau of Parks and Forestry of Guangzhou	Ecology as Priority, People-Centred, Urban–Rural Integration
		广州市花城绿城水城建设方案	Construction Plan of Guangzhou for Flower City, Green City and Water City	2014	2016	strategic plan	Municipal Government of Guangzhou	Committee of Urban–Rural Construction of Guangzhou	Scientific Development & Ecological Civilisation, & People-Centred
		广州市林业和园林十三五发展规划	The 13th Five-Year Plan for Forestry and Gardening Development in Guangzhou	2016	2020	action plan	Municipal Bureau of Parks and Forestry of Guangzhou	Municipal Bureau of Parks and Forestry of Guangzhou	Innovation, Coordination, Green, Open, Sharing, Scientific Development

(continued)

(continued)

Level	Theme	Title in Chinese	Title in English	Time of publication	Target period	Type	Publication entity	Coordination entity	Overarching political philosophy/policy principle
		广州市海绵城市专项规划	Special Plan for Sponge City of Guangzhou	2016	2030	special plan, technical guideline	Land Resources and Planning Commission of Guangzhou	Land Resources and Planning Commission of Guangzhou	Ecology as Priority, Ecological Civilisation
		广州市生态文明建设规划纲要	Outline of Ecological Civilization Construction in Guangzhou	2016	2020	strategic plan	Municipal Government of Guangzhou	All government institutions at or under municipal level in Guangzhou	Innovation, Coordination, Green, Open, Sharing, Scientific Development & Ecological Civilisation
	Urban spatial planning and development	广州市城市环境总体规划 2014–2030	Environmental Master Plan of Guangzhou City 2014–2030	2017	2030	action plan	Municipal Government of Guangzhou	Environmental Protection Bureau of Guangzhou	Innovation, Coordination, Green, Open, Sharing, Scientific Development, Environment as Priority

(continued)

(continued)

Level	Theme	Title in Chinese	Title in English	Time of publication	Target period	Type	Publication entity	Coordination entity	Overarching political philosophy/policy principle
		广州市公园建设与保护专项规划	Special Plan of Park Construction and Protection in Guangzhou	2017	2035	action plan	Municipal Bureau of Parks and Forestry of Guangzhou & Landscape Architecture Planning and Design Institute of Guangzhou	Municipal Bureau of Parks and Forestry of Guangzhou	Resource-Saving, Human-Nature Harmony
		广州市生态廊道总体规划与生态廊道规划建设指引	Ecological Corridor Master Plan and Construction Guideline of Guangzhou	2018	/	conceptual plan, technical guideline	Center of Urban Planning Research of Guangzhou	Municipal Bureau of Parks and Forestry of Guangzhou	Ecology as Priority, People-Centred
		广州市生态廊道地区项目建设指引	Ecological Corridor Regional Construction Guideline of Guangzhou	2018	/	technical guideline	Center of Urban Planning Research of Guangzhou	Municipal Bureau of Parks and Forestry of Guangzhou	/
		广州市生态廊道地区规划建设项目行政审批工作手册	Manual for Administrative Examination and Approval of the Ecological Corridor Regional Planning and Construction in Guangzhou	2018	/	administrative guideline	Center of Urban Planning Research of Guangzhou	Center of Urban Planning Research of Guangzhou	Ecological Civilisation

(continued)

(continued)

Level	Theme	Title in Chinese	Title in English	Time of publication	Target period	Type	Publication entity	Coordination entity	Overarching political philosophy/policy principle
		广州市生态廊道保护利用研究报告	Assessment Report on the Protection and Utilization of Ecological Corridors in Guangzhou	2018	/	report	Center of Urban Planning Research of Guangzhou	/	/
	Macro socio-economic plans	广州市经济和社会发展第十个五年计划	10th Five-Year Municipal Socio-Economic Development Plan of Guangzhou	2001	2005	macro socio-economic development plan	Development and Reform Commission of Guangzhou	All government institutions at or under municipal level in Guangzhou	Scientific Development, Sustainable Development
		广州市经济和社会发展第十一个五年计划	11th Five-Year Municipal Socio-Economic Development Plan of Guangzhou	2006	2010	macro socio-economic development plan	Development and Reform Commission of Guangzhou	All government institutions at or under municipal level in Guangzhou	Scientific Development, People-Centred, Harmonious Society
		广州市经济和社会发展第十二个五年计划	12th Five-Year Municipal Socio-Economic Development Plan of Guangzhou	2011	2015	macro socio-economic development plan	Development and Reform Commission of Guangzhou	All government institutions at or under municipal level in Guangzhou	Scientific Development
		广州市经济和社会发展第十三个五年计划	13th Five-Year Municipal Socio-Economic Development Plan of Guangzhou	2016	2020	macro socio-economic development plan	Development and Reform Commission of Guangzhou	All government institutions at or under municipal level in Guangzhou	Scientific Development, People-Centred, Sustainable Development

(continued)

(continued)

Level	Theme	Title in Chinese	Title in English	Time of publication	Target period	Type	Publication entity	Coordination entity	Overarching political philosophy/policy principle
	Work Report	广州市政府工作报告	Report on the Work of the Municipal Government of Guangzhou	2001	/	Report	Municipal Government of Guangzhou	/	Scientific Development, People-Centred, Sustainable Development
		广州市政府工作报告	Report on the Work of the Municipal Government of Guangzhou	2002	/	Report	Municipal Government of Guangzhou	/	Scientific Development, People-Centred, Sustainable Development
		广州市政府工作报告	Report on the Work of the Municipal Government of Guangzhou	2003	/	Report	Municipal Government of Guangzhou	/	Scientific Development, People-Centred, Sustainable Development
		广州市政府工作报告	Report on the Work of the Municipal Government of Guangzhou	2004	/	Report	Municipal Government of Guangzhou	/	Scientific Development, People-Centred, Sustainable Development
		广州市政府工作报告	Report on the Work of the Municipal Government of Guangzhou	2005	/	Report	Municipal Government of Guangzhou	/	Scientific Development, People-Centred, Sustainable Development

(continued)

(continued)

Level	Theme	Title in Chinese	Title in English	Time of publication	Target period	Type	Publication entity	Coordination entity	Overarching political philosophy/policy principle
		广州市政府工作报告	Report on the Work of the Municipal Government of Guangzhou	2006	/	Report	Municipal Government of Guangzhou	/	Scientific Development, People-Centred, Sustainable Development
		广州市政府工作报告	Report on the Work of the Municipal Government of Guangzhou	2007	/	Report	Municipal Government of Guangzhou	/	Scientific Development, People-Centred, Sustainable Development
		广州市政府工作报告	Report on the Work of the Municipal Government of Guangzhou	2008	/	Report	Municipal Government of Guangzhou	/	Scientific Development, People-Centred, Sustainable Development
		广州市政府工作报告	Report on the Work of the Municipal Government of Guangzhou	2009	/	Report	Municipal Government of Guangzhou	/	Scientific Development, People-Centred, Sustainable Development
		广州市政府工作报告	Report on the Work of the Municipal Government of Guangzhou	2010	/	Report	Municipal Government of Guangzhou	/	Scientific Development, People-Centred, Sustainable Development

(continued)

(continued)

Level	Theme	Title in Chinese	Title in English	Time of publication	Target period	Type	Publication entity	Coordination entity	Overarching political philosophy/policy principle
		广州市政府工作报告	Report on the Work of the Municipal Government of Guangzhou	2011	/	Report	Municipal Government of Guangzhou	/	Scientific Development, People-Centred, Sustainable Development
		广州市政府工作报告	Report on the Work of the Municipal Government of Guangzhou	2012	/	Report	Municipal Government of Guangzhou	/	Scientific Development, People-Centred, Sustainable Development
		广州市政府工作报告	Report on the Work of the Municipal Government of Guangzhou	2013	/	Report	Municipal Government of Guangzhou	/	Scientific Development, People-Centred, Sustainable Development
		广州市政府工作报告	Report on the Work of the Municipal Government of Guangzhou	2014	/	Report	Municipal Government of Guangzhou	/	Scientific Development, People-Centred, Sustainable Development
		广州市政府工作报告	Report on the Work of the Municipal Government of Guangzhou	2015	/	Report	Municipal Government of Guangzhou	/	Scientific Development, People-Centred, Sustainable Development

(continued)

(continued)

Level	Theme	Title in Chinese	Title in English	Time of publication	Target period	Type	Publication entity	Coordination entity	Overarching political philosophy/policy principle
		广州市政府工作报告	Report on the Work of the Municipal Government of Guangzhou	2016	/	Report	Municipal Government of Guangzhou	/	Scientific Development, People-Centred, Sustainable Development
		广州市政府工作报告	Report on the Work of the Municipal Government of Guangzhou	2017	/	Report	Municipal Government of Guangzhou	/	Scientific Development, People-Centred, Sustainable Development
		广州市政府工作报告	Report on the Work of the Municipal Government of Guangzhou	2018	/	Report	Municipal Government of Guangzhou	/	Scientific Development, People-Centred, Sustainable Development, Ecological Civilisation
		广州市政府工作报告	Report on the Work of the Municipal Government of Guangzhou	2019	/	Report	Municipal Government of Guangzhou	/	Scientific Development, People-Centred, Sustainable Development, Ecological Civilisation
Sub-municipal/ District level	Work Report	番禺区政府工作报告	Report on the Work of the District Government of Panyu	2006	/	Report	District Government of Panyu	/	Scientific Development, Harmonious Society

(continued)

(continued)

Level	Theme	Title in Chinese	Title in English	Time of publication	Target period	Type	Publication entity	Coordination entity	Overarching political philosophy/policy principle
		番禺区政府工作报告	Report on the Work of the District Government of Panyu	2007	/	Report	District Government of Panyu	/	Scientific Development, Harmonious Society
		番禺区政府工作报告	Report on the Work of the District Government of Panyu	2008	/	Report	District Government of Panyu	/	Scientific Development, Harmonious Society
		番禺区政府工作报告	Report on the Work of the District Government of Panyu	2009	/	Report	District Government of Panyu	/	Scientific Development, Harmonious Society
		番禺区政府工作报告	Report on the Work of the District Government of Panyu	2010	/	Report	District Government of Panyu	/	Scientific Development, Harmonious Society
		番禺区政府工作报告	Report on the Work of the District Government of Panyu	2011	/	Report	District Government of Panyu	/	Scientific Development
		番禺区政府工作报告	Report on the Work of the District Government of Panyu	2012	/	Report	District Government of Panyu	/	Scientific Development
		番禺区政府工作报告	Report on the Work of the District Government of Panyu	2013	/	Report	District Government of Panyu	/	Scientific Development

(continued)

(continued)

Level	Theme	Title in Chinese	Title in English	Time of publication	Target period	Type	Publication entity	Coordination entity	Overarching political philosophy/policy principle
	Work Report	海珠区政府工作报告	Report on the Work of the District Government of Haizhu	2009	/	Report	District Government of Haizhu	/	Scientific Development
		海珠区政府工作报告	Report on the Work of the District Government of Haizhu	2010	/	Report	District Government of Haizhu	/	Scientific Development
		海珠区政府工作报告	Report on the Work of the District Government of Haizhu	2011	/	Report	District Government of Haizhu	/	Scientific Development
		海珠区政府工作报告	Report on the Work of the District Government of Haizhu	2012	/	Report	District Government of Haizhu	/	Scientific Development, New Urbanisation
		海珠区政府工作报告	Report on the Work of the District Government of Haizhu	2013	/	Report	District Government of Haizhu	/	Scientific Development, New Urbanisation
		海珠区政府工作报告	Report on the Work of the District Government of Haizhu	2014	/	Report	District Government of Haizhu	/	Scientific Development, New Urbanisation
		海珠区政府工作报告	Report on the Work of the District Government of Haizhu	2015	/	Report	District Government of Haizhu	/	Reform & Innovation

(continued)

(continued)

Level	Theme	Title in Chinese	Title in English	Time of publication	Target period	Type	Publication entity	Coordination entity	Overarching political philosophy/policy principle
		海珠区政府工作报告	Report on the Work of the District Government of Haizhu	2016	/	Report	District Government of Haizhu	/	Scientific Development, People-Centred, Innovation, Coordination, Green, Open, Sharing, Sustainable Development
	Work Report	天河区政府工作报告	Report on the Work of the District Government of Tianhe	2014	/	Report	District Government of Tianhe	/	/
		天河区政府工作报告	Report on the Work of the District Government of Tianhe	2015	/	Report	District Government of Tianhe	/	/
		天河区政府工作报告	Report on the Work of the District Government of Tianhe	2016	/	Report	District Government of Tianhe	/	/
		天河区政府工作报告	Report on the Work of the District Government of Tianhe	2017	/8	Report	District Government of Tianhe	/	New Development
		天河区政府工作报告	Report on the Work of the District Government of Tianhe	2018	/	Report	District Government of Tianhe	/	New Development
		天河区政府工作报告	Report on the Work of the District Government of Tianhe	2019	/	Report	District Government of Tianhe	/	New Development

List of Scientific Publications During the PhD

- Ferreira, Lurdes J., Luís P. Dias, and **Jieling Liu.** 2022. Adopting Carbon Pricing Tools at the Local Level: A City Case Study in Portugal. *Sustainability* 14(3):1812. https://doi.org/10.3390/su14031812
- **Jieling Liu,** Franz Gatzweiler, Simon Hodson, Gabriele Harrer-Puchner, Giles B. Sioen, Mamello Thinyane, Ronit Purian, Virginia Murray, Xiaofeng Yi & Alejandro Camprubi. 2022. Co-creating solutions to complex urban problems with collaborative systems modelling - insights from a workshop on health co-benefits of urban green spaces in Guangzhou. *Cities & Health*, DOI: https://doi.org/10.1080/23748834.2022.2026694
- **Jieling Liu & Franz Gatzweiler.2020**. The Institutional Challenge to Co-Deliver Migrant Integration and Urban Greening—Evidence from Haizhu Wetland Park Project in Guangzhou, China. *Journal of Chinese Governance.* DOI: https://doi.org/10.1080/23812346.2020.1760069
- **Jieling Liu.** 2021. COVID-19 Shows Us the Need to Plan Urban green spaces More Systematically for Urban Health and Wellbeing. In: Gatzweiler F. (eds) *Urban Health and Wellbeing Programme. Urban Health and Wellbeing.* Springer, Singapore. Zhejiang University Press. https://doi.org/10.1007/978-981-33-6036-5_7. ISBN: 978-981-336-036-5.
- **Jieling Liu.** 2020. Institutional Barriers and Transformation to Climate Change Management in Emerging Coastal Megacities: A Case Study from Guangzhou, China. Chapter of book: *Handbook of Climate Change Management.* DOI: https://doi.org/10.1007/978-3-030-22759-3_37-1. Part of ISBN: 9783030227593
- **Jieling Liu,** Franz Gatzweiler, Manasi Kumar. 2020. An Evolutionary Complex Systems Perspective on Urban Health. *Socio-Economic Planning Sciences.* 2020. DOI: https://doi.org/10.1016/j.seps.2020.100815
- **Jieling Liu**. 2020. Narrating the Impacts of Climate Change for Urban Health Governance in Guangzhou, China. *Journal of Cities and Health.* Published online: January 2020. DOI: https://doi.org/10.1080/23748834.2019.1701945
- Thomas Elmqvist, Franz Gatzweiler, Elisabet Lindgren, **Jieling Liu**. 2019. Resilience Management for Healthy Cities in a Changing Climate. Chapter of book: Marselle M., Stadler J., Korn H., Irvine K., Bonn A. (eds) *Biodiversity and Health in the Face of Climate Change.* Springer, Cham. First Online: June 2019. https://doi.org/10.1007/978-3-030-02318-8_18
- **Jieling Liu**. 2019. Financing and Implementing Resilience with a Systems Approach in Beirut. In: Gatzweiler F. (eds) *Urban Health and Wellbeing Programme. Urban Health and Wellbeing (Systems Approaches)*. Springer, Singapore. https://doi.org/10.1007/978-981-15-1380-0_9. ISBN: 978-981-15-1379-4.
- **Jieling Liu,** Franz Gatzweiler, Luísa Schmidt. 2019. Addressing the Nexus Challenge of Climate Change, Human Health, and Urban Sustainability. *ICS Estudos e Relatórios*. ISSN: 2183-6922. Available at: https://www.ics.ulisboa.pt/flipping/er2019_1/12/

- Charles Ebikeme, Franz Gatzweiler, Tolullah Oni, **Jieling Liu**, Andrea Oyuela & José Siri. 2019. Xiamen Call for Action: Building the Brain of the City—Universal Principles of Urban Health, *Journal of Urban Health.* Published online: January 2019. Springer: Volume 96, Issue 4, pp 507-509. ©The New York Academy of Medicine. DOI: https://doi.org/10.1007/s11524-018-00342-0

Declaration of Original Authorship

I certify that the book is re-edited from my original PhD thesis for the PhD degree of Climate Change and Sustainable Development Policies from at the Institute of Social Sciences—University of Lisbon, and is solely my own work other than where I have clearly indicated that it is the work of others (in which case the extent of any work carried out jointly by me and any other person is clearly identified in it).

The copyright of this book rests with the author. Quotation from it is permitted, provided that full acknowledgement is made. This book may not be reproduced without my prior written consent.

I warrant that this authorization does not, to the best of my belief, infringe the rights of any third party.

I declare that my book consists of 93,028 words of which 66,203 are in the main text and the rest are in the bibliography and appendices.

Jieling Liu

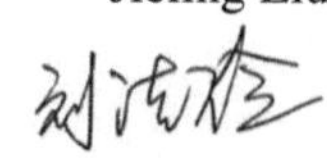

June 22, 2022
With most recent edit on May 18, 2023

Uncited References

2018 Yearbook of Meteorological Disasters in China (2018). https://www.yearbookchina.com/navibooklist-n3020041310-1.html

Alberti M, Marzluff JM, Shulenberger E, Bradley G, Ryan C, Zumbrunnen C (2008) Integrating humans into ecology: Opportunities and challenges for studying urban ecosystems. In Urban ecology: an international perspective on the interaction between humans and nature. https://doi.org/10.1007/978-0-387-73412-5_9

Bai X, Chen J, Shi P (2012) Landscape urbanization and economic growth in China: Positive feedbacks and sustainability dilemmas. Environ Sci Technol. https://doi.org/10.1021/es202329f

Bowler DE, Buyung-Ali L, Knight TM, Pullin AS (2010) Urban greening to cool towns and cities: a systematic review of the empirical evidence. Landsc Urban Plan. https://doi.org/10.1016/j.landurbplan.2010.05.006

Burgess EW (n.d.) The growth of the city: an introduction to a research project. In: Urban ecology. Springer, US, pp 71–78. https://doi.org/10.1007/978-0-387-73412-5_5

Elmqvist T, Goodness J, Marcotullio PJ, Parnell S, Sendstad M, Wilkinson C, Fragkias M, Güneralp B, McDonald RI, Schewenius M, Seto KC (2013a) Urbanization, biodiversity and ecosystem services: challenges and opportunities: a global assessment. In: Urbanization, biodiversity and ecosystem services: challenges and opportunities: a global assessment. https://doi.org/10.1007/978-94-007-7088-1

Field CB, Barros VR, Mastrandrea MD, Mach KJ, Abdrabo MA-K, Adger WN, Anokhin YA, Anisimov OA, Arent DJ, Barnett J, Burkett VR, Cai R, Chatterjee M, Cohen SJ, Cramer W, Dasgupta Purnamita,... (2014) Climate change 2014: impacts, adaptation, and vulnerability—summary for policymakers. Cambridge University Press. https://www.ipcc.ch/site/assets/uploads/2018/02/ar5_wgII_spm_en.pdf

Grimm NB, Grove JM, Pickett STA, Redman CL (n.d.) Integrated Approaches to Long-Term Studies of Urban Ecological Systems. In *Urban Ecology* (pp. 123–141). Springer US. https://doi.org/10.1007/978-0-387-73412-5_8

Guan D, Chen Y (2003) Status of urban vegetation in Guangzhou City. J Forest Res 14(3):249–252. https://doi.org/10.1007/bf02856840

Hardin PJ, Jensen RR (2007) The effect of urban leaf area on summertime urban surface kinetic temperatures: a Terre Haute case study. Urban Forest Urban Green. https://doi.org/10.1016/j.ufug.2007.01.005

© Zhejiang University Press 2023

J. Liu, *Transforming Urban Green Space Governance in China Under Ecological Civilization: An Institutional Analysis*, Urban Health and Wellbeing,
https://doi.org/10.1007/978-981-99-6694-3

Hillsdon M, Panter J, Foster C, Jones A (2006) The relationship between access and quality of urban green space with population physical activity. Public Health. https://doi.org/10.1016/j.puhe.2006.10.007

Knoblauch D, Naumann S, Mederake L, Sosa A. (n.d.). Gaps and opportunities to sustainable urban development through nature-based solutions . Retrieved November 17, 2020, from H2020 grant no. 776604

Lawrence RJ, Gatzweiler FW (2017) Wanted : a transdisciplinary knowledge domain for urban health. https://doi.org/10.1007/s11524-017-0182-x

Li Z, Wang L, Guan D (2000) Landscape heterogeneity of urban vegetation in Guangzhou (in Chinese: 广州城市绿地系统景观异质性分析). J Appl Ecol (in Chinese: 应用生态学报), 11(1):127–130.

Li Z (ed) (2016) Research on the theory, methods and practice of public policy evaluation, 2nd edn. China Development Press

Liu X, Huang J, Zhu J (2019) Property-rights regime in transition: understanding the urban regeneration process in China—A case study of Jinhuajie, Guangzhou. Cities. https://doi.org/10.1016/j.cities.2019.02.008

Meadows DH, Meadows DL, Randers J, Behrens WW (1972) The limits to growth : a report to the club of Rome. Universe 1–9. https://doi.org/10.1080/07293682.2004.9982332

National new-type urbanisation plan (2014–2020) 国家新型城镇化规划 (2014–2020年) (2014) (testimony of State Council of China). http://www.gov.cn/zhengce/2014-03/16/content_2640075.htm

National new-type urbanization plan 2014–2020 (2014) http://www.gov.cn/zhengce/2014-03/16/content_2640075.htm

North DC, Douglass C (1990) Institutions, institutional change, and economic performance. Cambridge University Press

Renn O (2008) Risk governance: coping with uncertainty in a complex World. Earthscan

Smith A (1776) Of the origin and use of money. In: Wealth of nations, pp 25–30. http://geolib.com/smith.adam/won1-04.html

State Council of China (2014) National Disaster Reduction Center of the Ministry of Civil Affairs released information about the basic situation of natural disasters nationwide 2013–2017 (in Chinese: 民政部减灾办发布2013年全国自然灾害基本情况)

Sukopp H (n.d.) On the early history of urban ecology in Europe. In: Urban ecology. Springer US, pp 79–97. https://doi.org/10.1007/978-0-387-73412-5_6

Takano T, Nakamura K, Watanabe M (2002) Urban residential environments and senior citizens' longevity in megacity areas: The importance of walkable green spaces. J Epidemiol Community Health. https://doi.org/10.1136/jech.56.12.913

United Nations Development Programme (2020) United nations sustainable development cooperation framework: People's Republic of China 2021–2025, 1st edn. United Nations

United Nations Office for Disaster Risk Reduction (2015) Sendai framework for disaster risk reduction 2015–2030. https://www.preventionweb.net/files/43291_sendaiframeworkfordrren.pdf

Wang HY, Rosenau JN (2009) China and global governance. Asian Perspect 33(3):5–39

World Health Organization. (n.d.). *Physical Inactivity: A Global Public Health Problem.* Global Strategy on Diet, Physical Activity and Health.

World Health Organization (2015) WHO workplan on climate change and health aims and objectives: 2014–2019. https://www.who.int/docs/default-source/climate-change/who-workplan-on-climate-change-and-health-2014---2019.pdf?sfvrsn=501ed812_2

World Health Organization (2018a) COP24 special report: health and climate change.

Xia J, Zhang YY, Xiong LH, He S, Wang LF, Yu ZB (2017) Opportunities and challenges of the Sponge City construction related to urban water issues in China. Sci China Earth Sci. https://doi.org/10.1007/s11430-016-0111-8

Yao-Dong D, Xian-Wei W, Xiao-Feng Y, Wen-Jun M, Hui A, Xiao-Xuan W (2013) Impacts of climate change on human health and adaptation strategies in South China. Adv Clim Chang Res. https://doi.org/10.3724/SP.J.1248.2013.208

Zhu Da, Ru J (2008) Strategic environmental assessment in China: Motivations, politics, and effectiveness. J Environ Manage. https://doi.org/10.1016/j.jenvman.2007.03.040

MIX
Papier aus verantwortungsvollen Quellen
Paper from responsible sources
FSC® C105338

If you have any concerns about our products,
you can contact us on
ProductSafety@springernature.com

In case Publisher is established outside the EU,
the EU authorized representative is:
Springer Nature Customer Service Center GmbH
Europaplatz 3, 69115 Heidelberg, Germany

Printed by Libri Plureos GmbH
in Hamburg, Germany